cram101 Textbook Outlines

E-TEXTBOOK VERSION

Outlines, Notes & Highlights for:

Cram101 Textbook Reviews

Just The facts101

Textbook Key Facts

e-Study Guide

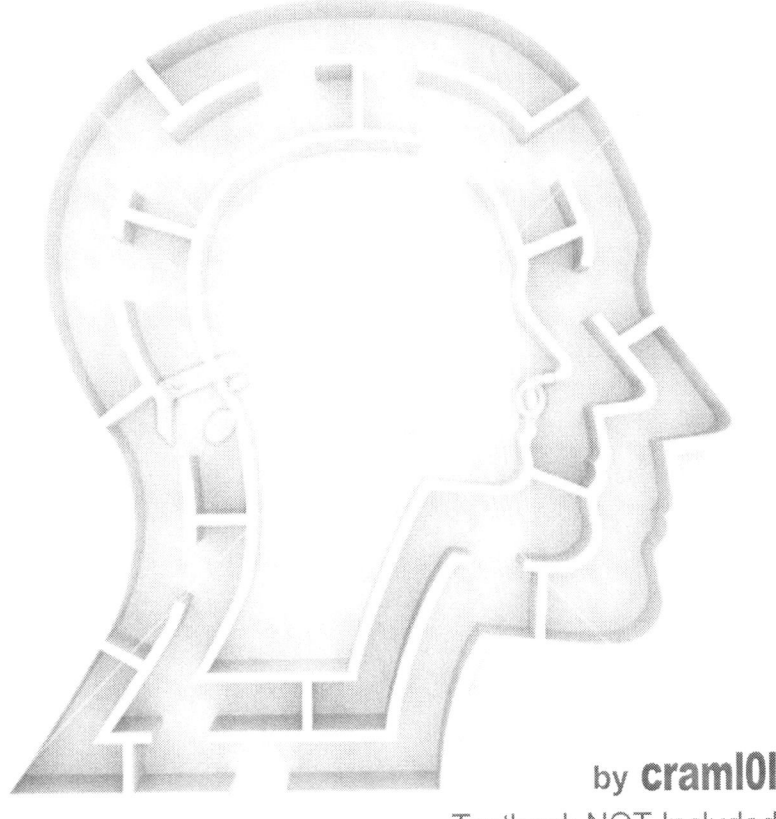

by **cram101**
Textbook NOT Included

Textbook Outlines, Highlights, and Practice Quizzes

Chemistry for the Biosciences: The Essential Concepts

by Jonathan Crowe, 2nd Edition

All "Just the Facts101" Material Written or Prepared by Cram101 Publishing

Title Page

WHY STOP HERE... THERE'S MORE ONLINE

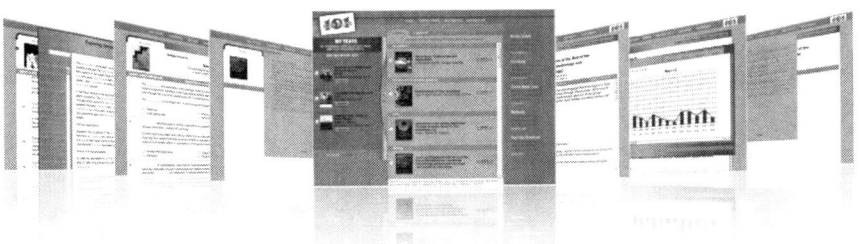

With technology and experience, we've developed tools that make studying easier and efficient. Like this Cram101 textbook notebook, **Cram101.com** offers you the highlights from every chapter of your actual textbook. However, unlike this notebook, **Cram101.com** gives you practice tests for each of the chapters. You also get access to in-depth reference material for writing essays and papers.

By purchasing this book, you get 50% off the normal subscription free!. Just enter the promotional code **'DK73DW19910'** on the Cram101.com registration screen.

CRAM101.COM FEATURES:

Outlines & Highlights
Just like the ones in this notebook, but with links to additional information.

Integrated Note Taking
Add your class notes to the Cram101 notes, print them and maximize your study time.

Problem Solving
Step-by-step walk throughs for math, stats and other disciplines.

Practice Exams
Five different test taking formats for every chapter.

Easy Access
Study any of your books, on any computer, anywhere.

Unlimited Textbooks
All the features above for virtually all your textbooks, just add them to your account at no additional cost.

Be sure to use the promo code above when registering on Cram101.com to get 50% off your membership fees.

STUDYING MADE EASY

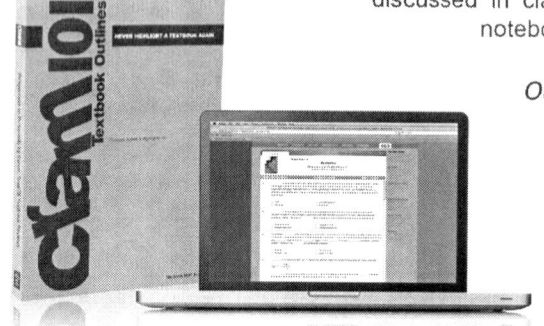

This Cram101 notebook is designed to make studying easier and increase your comprehension of the textbook material. Instead of starting with a blank notebook and trying to write down everything discussed in class lectures, you can use this Cram101 textbook notebook and annotate your notes along with the lecture.

Our goal is to give you the best tools for success.

For a supreme understanding of the course, pair your notebook with our online tools. Should you decide you prefer Cram101.com as your study tool,

we'd like to offer you a trade...

Our Trade In program is a simple way for us to keep our promise and provide you the best studying tools, regardless of where you purchased your Cram101 textbook notebook. As long as your notebook is in *Like New Condition**, you can send it back to us and we will immediately give you a Cram101.com account free for 120 days!

Let The Trade In Begin!

THREE SIMPLE STEPS TO TRADE:

1. Go to www.cram101.com/tradein and fill out the packing slip information.
2. Submit and print the packing slip and mail it in with your Cram101 textbook notebook.
3. Activate your account after you receive your email confirmation.

* Books must be returned in *Like New Condition*, meaning there is no damage to the book including, but not limited to; ripped or torn pages, markings or writing on pages, or folded / creased pages. Upon receiving the book, Cram101 will inspect it and reserves the right to terminate your free Cram101.com account and return your textbook notebook at the owners expense.

"Just the Facts101" is a Cram101 publication and tool designed to give you all the facts from your textbooks. Visit Cram101.com for the full practice test for each of your chapters for virtually any of your textbooks.

Cram101 has built custom study tools specific to your textbook. We provide all of the factual testable information and unlike traditional study guides, we will never send you back to your textbook for more information.

YOU WILL NEVER HAVE TO HIGHLIGHT A BOOK AGAIN!

Cram101 StudyGuides
All of the information in this StudyGuide is written specifically for your textbook. We include the key terms, places, people, and concepts... the information you can expect on your next exam!

Want to take a practice test?
Throughout each chapter of this StudyGuide you will find links to cram101.com where you can select specific chapters to take a complete test on, or you can subscribe and get practice tests for up to 12 of your textbooks, along with other exclusive cram101.com tools like problem solving labs and reference libraries.

Cram101.com
Only cram101.com gives you the outlines, highlights, and PRACTICE TESTS specific to your textbook. Cram101.com is an online application where you'll discover study tools designed to make the most of your limited study time.

By purchasing this book, you get 50% off the normal subscription free!. Just enter the promotional code **'DK73DW19910'** on the Cram101.com registration screen.

www.Cram101.com

Copyright © 2012 by Cram101, Inc. All rights reserved.
"Just the FACTS101"®, "Cram101"® and "Never Highlight a Book Again!"® are registered trademarks of Cram101, Inc.
ISBN(s): 9781478407508. PUBl-6.201279

Learning System

facts101

Chemistry for the Biosciences: The Essential Concepts
Jonathan Crowe, 2nd

CONTENTS

1. Introduction: why biologists need chemistry 5
2. Atoms: the foundations of life 18
3. Compounds and chemical bonding: bringing atoms together 33
4. Molecular interactions: holding it all together 46
5. Organic compounds 1: the framework of life 62
6. Organic compounds 2: adding function to the framework of life 77
7. Biological macromolecules: providing life's infrastructure 98
8. Molecular shape and structure 1: from atoms to small molecules 107
9. Molecular shape and structure 2: the shape of large molecules 120
10. Isomerism: generating chemical variety 131
11. Chemical analysis 1: how do we know what is there? 152
12. Chemical analysis 2: how do we know how much is there? 168
13. Energy: what makes reactions go? 184
14. Kinetics: what affects the speed of a reaction? 196
15. Equilibria: how far do reactions go? 206
16. Acids, bases, and the aqueous environment: the medium of life 217
17. Chemical reactions 1: bringing molecules to life 228
18. Chemical reactions 2: reaction mechanisms driving the chemistry of life 247

CHAPTER OUTLINE: KEY TERMS, PEOPLE, PLACES, CONCEPTS
Chapter 1
Introduction: why biologists need chemistry

- _____ ATP synthase
- _____ Mitochondrion
- _____ DNA-binding protein
- _____ DNA profiling
- _____ Amino acid
- _____ Nucleic acid
- _____ Chemical nomenclature

CHAPTER HIGHLIGHTS: KEY TERMS, PEOPLE, PLACES, CONCEPTS
Chapter 1. Introduction: why biologists need chemistry

ATP synthase	ATP synthase is an important enzyme that provides energy for the cell to use through the synthesis of adenosine triphosphate (ATP). ATP is the most commonly used 'energy currency' of cells from most organisms. It is formed from adenosine diphosphate (ADP) and inorganic phosphate (P_i), and needs energy.
Mitochondrion	In cell biology, a mitochondrion is a membrane-enclosed organelle found in most eukaryotic cells. These organelles range from 0.5 to 1.0 micrometer (μm) in diameter. Mitochondria are sometimes described as 'cellular power plants' because they generate most of the cell's supply of adenosine triphosphate (ATP), used as a source of chemical energy.
DNA-binding protein	DNA-binding proteins are proteins that are composed of DNA-binding domains and thus have a specific or general affinity for either single or double stranded DNA. Sequence-specific DNA-binding proteins generally interact with the major groove of B-DNA, because it exposes more functional groups that identify a base pair. However there are some known minor groove DNA-binding ligands such as Netropsin, Distamycin, Hoechst 33258, Pentamidine and others. Examples DNA-binding proteins include transcription factors which modulate the process of transcription, various polymerases, nucleases which cleave DNA molecules, and histones which are involved in chromosome packaging and transcription in the cell nucleus.
DNA profiling	DNA profiling is a technique employed by forensic scientists to assist in the identification of individuals by their respective DNA profiles. DNA profiles are encrypted sets of numbers that reflect a person's DNA makeup, which can also be used as the person's identifier. DNA profiling should not be confused with full genome sequencing.
Amino acid	Amino acids (?'mi?no?, ?'ma?o?, or 'æm?o?) are molecules containing an amine group, a carboxylic acid group, and a side-chain that is specific to each amino acid. The key elements of an amino acid are carbon, hydrogen, oxygen, and nitrogen. They are particularly important in biochemistry, where the term usually refers to alpha-amino acids.
Nucleic acid	Nucleic acids are biological molecules essential for known forms of life on this planet; they include DNA (deoxyribonucleic acid) and RNA (ribonucleic acid). Together with proteins, nucleic acids are the most important biological macromolecules; each is found in abundance in all living things, where they function in encoding, transmitting and expressing genetic information.

Chapter 1. Introduction: why biologists need chemistry

Nucleic acids were discovered by Friedrich Miescher in 1869. Experimental studies of nucleic acids constitute a major part of modern biological and medical research, and form a foundation for genome and forensic science, as well as the biotechnology and pharmaceutical industries.

Chemical nomenclature	A chemical nomenclature is a set of rules to generate systematic names for chemical compounds. The nomenclature used most frequently worldwide is the one created and developed by the International Union of Pure and Applied Chemistry (IUPAC).
	The IUPAC's rules for naming organic and inorganic compounds are contained in two publications, known as the Blue Book and the Red Book, respectively.

PRACTICE QUIZ
Chapter 1. Introduction: why biologists need chemistry

1. _____ is an important enzyme that provides energy for the cell to use through the synthesis of adenosine triphosphate (ATP). ATP is the most commonly used 'energy currency' of cells from most organisms. It is formed from adenosine diphosphate (ADP) and inorganic phosphate (P_i), and needs energy.

    a. Ecarin
    b. Endoenzyme
    c. ATP synthase
    d. Endoglycosidase H

2. _____ is a technique employed by forensic scientists to assist in the identification of individuals by their respective DNA profiles. DNA profiles are encrypted sets of numbers that reflect a person's DNA makeup, which can also be used as the person's identifier. _____ should not be confused with full genome sequencing.

    a. DNA sequencer
    b. Peak calling
    c. DNA supercoil
    d. DNA profiling

3. A _____ is a set of rules to generate systematic names for chemical compounds. The nomenclature used most frequently worldwide is the one created and developed by the International Union of Pure and Applied Chemistry (IUPAC).

    The IUPAC's rules for naming organic and inorganic compounds are contained in two publications, known as the Blue Book and the Red Book, respectively.

    a. Line notation
    b. Molecular entity
    c. Chemical nomenclature
    d. Noble metal

4. In cell biology, a _____ is a membrane-enclosed organelle found in most eukaryotic cells. These organelles range from 0.5 to 1.0 micrometer (μm) in diameter. Mitochondria are sometimes described as 'cellular power plants' because they generate most of the cell's supply of adenosine triphosphate (ATP), used as a source of chemical energy.

---

a. NADH dehydrogenase
   b. Nanaerobe
   c. Mitochondrion
   d. P/O ratio

5. _____s are proteins that are composed of DNA-binding domains and thus have a specific or general affinity for either single or double stranded DNA. Sequence-specific _____s generally interact with the major groove of B-DNA, because it exposes more functional groups that identify a base pair. However there are some known minor groove DNA-binding ligands such as Netropsin, Distamycin, Hoechst 33258, Pentamidine and others.

   Examples
   _____s include transcription factors which modulate the process of transcription, various polymerases, nucleases which cleave DNA molecules, and histones which are involved in chromosome packaging and transcription in the cell nucleus.

   a. Flory convention
   b. DNA-binding protein
   c. FlyFactorSurvey
   d. GHK flux equation

ANSWER KEY
Chapter 1. Introduction: why biologists need chemistry

1. c
2. d
3. c
4. c
5. b

You can take the complete Chapter Practice Test

for Chapter 1. Introduction: why biologists need chemistry
on all key terms, persons, places, and concepts.

Online 99 Cents

http://www.epub14.51.19910.1.cram101.com/

Use www.Cram101.com for all your study needs

including Cram101's online interactive problem solving labs in chemistry, statistics, mathematics, and more.

CHAPTER OUTLINE: KEY TERMS, PEOPLE, PLACES, CONCEPTS
Chapter 2
Atoms: the foundations of life

- Chemical symbol
- Group
- Carbon
- DNA profiling
- Atomic number
- Electron
- Neutron
- Subatomic particle
- Mass number
- Xenon
- Intramolecular force
- Isotope
- Photosynthesis
- Atomic weight
- Deuterium
- Atomic mass
- Beryllium
- Helium
- Atomic orbital

Chapter 2. Atoms: the foundations of life

	Diameter
	Gel electrophoresis
	Pauli exclusion principle
	Aufbau principle
	Covalent bond
	Degenerate orbital
	Chlorophyll
	Ground state
	Reversible reaction
	Disulfide bond
	Electromagnetic radiation
	Electromagnetic spectrum
	Racemic mixture
	Radio waves
	Wavelength
	Lewis structure
	Emission spectrum
	Electron pair
	Valence electron

CHAPTER HIGHLIGHTS: KEY TERMS, PEOPLE, PLACES, CONCEPTS
Chapter 2. Atoms: the foundations of life

Chemical symbol	A chemical symbol is a 1- or 2-letter internationally agreed code for a chemical element, usually derived from the name of the element, often in Latin.
	Only the first letter is capitalised. For example, 'He' is the symbol for helium, 'Pb' for lead, 'W' for tungsten.
Group	In chemistry, a group (also known as a family) is a vertical column in the periodic table of the chemical elements. There are 18 groups in the standard periodic table, including the d-block elements, but excluding the f-block elements.
	The explanation of the pattern of the table is that the elements in a group have similar physical or chemical characteristic of the outermost electron shells of their atoms (i.e. the same core charge), as most chemical properties are dominated by the orbital location of the outermost electron.
Carbon	Carbon 'k?rb?n is the chemical element with symbol C and atomic number 6. As a member of group 14 on the periodic table, it is nonmetallic and tetravalent--making four electrons available to form covalent chemical bonds. There are three naturally occurring isotopes, with ^{12}C and ^{13}C being stable, while ^{14}C is radioactive, decaying with a half-life of about 5,730 years. Carbon is one of the few elements known since antiquity.
DNA profiling	DNA profiling is a technique employed by forensic scientists to assist in the identification of individuals by their respective DNA profiles. DNA profiles are encrypted sets of numbers that reflect a person's DNA makeup, which can also be used as the person's identifier. DNA profiling should not be confused with full genome sequencing.
Atomic number	In chemistry and physics, the atomic number is the number of protons found in the nucleus of an atom and therefore identical to the charge number of the nucleus. It is conventionally represented by the symbol Z. The atomic number uniquely identifies a chemical element. In an atom of neutral charge, the atomic number is also equal to the number of electrons.

Chapter 2. Atoms: the foundations of life

Electron	The electron is a subatomic particle with a negative elementary electric charge. It has no known components or substructure; in other words, it is generally thought to be an elementary particle. An electron has a mass that is approximately 1/1836 that of the proton.
Neutron	The neutron is a subatomic hadron particle which has the symbol n or n^0, no net electric charge and a mass slightly larger than that of a proton. With the exception of hydrogen, nuclei of atoms consist of protons and neutrons, which are therefore collectively referred to as nucleons. The number of protons in a nucleus is the atomic number and defines the type of element the atom forms.
Subatomic particle	In physics or chemistry, subatomic particles are the particles, which are smaller than an atom. There are two types of subatomic particles: elementary particles, which are not made of other particles, and composite particles. Particle physics and nuclear physics study these particles and how they interact.
Mass number	The mass number also called atomic mass number, is the total number of protons and neutrons (together known as nucleons) in an atomic nucleus. Because protons and neutrons both are baryons, the mass number A is identical with the baryon number B as of the nucleus as of the whole atom or ion. The mass number is different for each different isotope of a chemical element.
Xenon	Xenon is a chemical element with the symbol Xe and atomic number 54. A colorless, heavy, odorless noble gas, xenon occurs in the Earth's atmosphere in trace amounts. Although generally unreactive, xenon can undergo a few chemical reactions such as the formation of xenon hexafluoroplatinate, the first noble gas compound to be synthesized. Naturally occurring xenon consists of eight stable isotopes.
Intramolecular force	An intramolecular force is any force that holds together the atoms making up a molecule or compound. They contain all types of chemical bond. They are stronger than intermolecular forces, which are present between atoms or molecules that are not actually bonded.
Isotope	Isotopes are variants of a particular chemical element. While all isotopes of a given element share the same number of protons, each isotope differs from the others in its number of neutrons. The term isotope is formed from the Greek roots isos (?σος 'equal') and topos (τ?πος 'place').

Chapter 2. Atoms: the foundations of life

Photosynthesis	Photosynthesis is a process used by plants and other organisms to capture the sun's energy to split off water's hydrogen from oxygen. Hydrogen is combined with carbon dioxide (absorbed from air or water) to form glucose and release oxygen. All living cells in turn use fuels derived from glucose and oxidize the hydrogen and carbon to release the sun's energy and reform water and carbon dioxide in the process (cellular respiration).
Atomic weight	Atomic weight is a dimensionless physical quantity, the ratio of the average mass of atoms of an element (from a given source) to 1/12 of the mass of an atom of carbon-12 (known as the unified atomic mass unit). The term is usually used, without further qualification, to refer to the standard atomic weights published at regular intervals by the International Union of Pure and Applied Chemistry (IUPAC) and which are intended to be applicable to normal laboratory materials.
Deuterium	Deuterium, is one of two stable isotopes of hydrogen. It has a natural abundance in Earth's oceans of about one atom in 6,420 of hydrogen (~156.25 ppm on an atom basis). Deuterium accounts for approximately 0.0156% of all naturally occurring hydrogen in Earth's oceans, while the most common isotope (hydrogen-1 or protium) accounts for more than 99.98%.
Atomic mass	The atomic mass is the mass of a specific isotope, most often expressed in unified atomic mass units. The atomic mass is the total mass of protons, neutrons and electrons in a single atom. The atomic mass is sometimes incorrectly used as a synonym of relative atomic mass, average atomic mass and atomic weight; these differ subtly from the atomic mass.
Beryllium	Beryllium is the chemical element with the symbol Be and atomic number 4. Because any beryllium synthesized in stars is short-lived, it is a relatively rare element in both the universe and in the crust of the Earth. It is a divalent element which occurs naturally only in combination with other elements in minerals. Notable gemstones which contain beryllium include beryl (aquamarine, emerald) and chrysoberyl.
Helium	

Chapter 2. Atoms: the foundations of life

	Helium is the chemical element with atomic number 2 and an atomic weight of 4.002602, which is represented by the symbol He. It is a colorless, odorless, tasteless, non-toxic, inert, monatomic gas that heads the noble gas group in the periodic table. Its boiling and melting points are the lowest among the elements and it exists only as a gas except in extreme conditions.
Atomic orbital	An atomic orbital is a mathematical function that describes the wave-like behavior of either one electron or a pair of electrons in an atom. This function can be used to calculate the probability of finding any electron of an atom in any specific region around the atom's nucleus. The term may also refer to the physical region where the electron can be calculated to be, as defined by the particular mathematical form of the orbital.
Diameter	In geometry, a diameter of a circle is any straight line segment that passes through the center of the circle and whose endpoints are on the circle. The diameters are the longest chords of the circle. The word 'diameter' derives from Greek δι?μετρος (diametros), 'diagonal of a circle', from δια- (dia-), 'across, through' + μ?τρον (metron), 'a measure').
Gel electrophoresis	Gel electrophoresis is a method used in clinical chemistry to separate proteins by charge and or size (IEF agarose, essentially size independent) and in biochemistry and molecular biology to separate a mixed population of DNA and RNA fragments by length, to estimate the size of DNA and RNA fragments or to separate proteins by charge. Nucleic acid molecules are separated by applying an electric field to move the negatively charged molecules through an agarose matrix. Shorter molecules move faster and migrate farther than longer ones because shorter molecules migrate more easily through the pores of the gel.
Pauli exclusion principle	The Pauli exclusion principle is the quantum mechanical principle that no two identical fermions (particles with half-integer spin) may occupy the same quantum state simultaneously. A more rigorous statement is that the total wave function for two identical fermions is anti-symmetric with respect to exchange of the particles. The principle was formulated by Austrian physicist Wolfgang Pauli in 1925.
Aufbau principle	The Aufbau principle is used to determine the electron configuration of an atom, molecule or ion. The principle postulates a hypothetical process in which an atom is 'built up' by progressively adding electrons. As they are added, they assume their most stable conditions (electron orbitals) with respect to the nucleus and those electrons already there.

Chapter 2. Atoms: the foundations of life

Covalent bond	A covalent bond is a form of chemical bonding that is characterized by the sharing of pairs of electrons between atoms. The stable balance of attractive and repulsive forces between atoms when they share electrons is known as covalent bonding.
	Covalent bonding includes many kinds of interaction, including σ-bonding, π-bonding, metal-to-metal bonding, agostic interactions, and three-center two-electron bonds.
Degenerate orbital	Degenerate orbitals for electrons in an atomic subshell are orbitals at identical energy levels. These are important in physical chemistry because they affect the ways electrons fill atoms. For example, all the 3p orbitals have same energy level, and so do all the 5d orbitals.
Chlorophyll	Chlorophyll is a green pigment found in almost all plants, algae, and cyanobacteria. Its name is derived from the Greek words χλωρος, chloros ('green') and φ?λλον, phyllon ('leaf'). Chlorophyll is an extremely important biomolecule, critical in photosynthesis, which allows plants to absorb energy from light.
Ground state	The ground state of a quantum mechanical system is its lowest-energy state; the energy of the ground state is known as the zero-point energy of the system. An excited state is any state with energy greater than the ground state. The ground state of a quantum field theory is usually called the vacuum state or the vacuum.
Reversible reaction	A reversible reaction is a chemical reaction that results in an equilibrium mixture of reactants and products. For a reaction involving two reactants and two products this can be expressed symbolically as $$aA + bB \rightleftharpoons cC + dD$$ A and B can react to form C and D or, in the reverse reaction, C and D can react to form A and B. This is distinct from reversible process in thermodynamics.

Chapter 2. Atoms: the foundations of life

	The concentrations of reactants and products in an equilibrium mixture are determined by the analytical concentrations of the reagents (A and B or C and D) and the equilibrium constant, K. The magnitude of the equilibrium constant depends on the Gibbs free energy change for the reaction.
Disulfide bond	In chemistry, a disulfide bond is a covalent bond, usually derived by the coupling of two thiol groups. The linkage is also called an SS-bond or disulfide bridge. The overall connectivity is therefore R-S-S-R. The terminology is widely used in biochemistry.
Electromagnetic radiation	Electromagnetic radiation is a form of energy emitted and absorbed by charged particles, which exhibits wave-like behavior as it travels through space. EMR has both electric and magnetic field components, which stand in a fixed ratio of intensity to each other, and which oscillate in phase perpendicular to each other and perpendicular to the direction of energy and wave propagation. In vacuum, electromagnetic radiation propagates at a characteristic speed, the speed of light.
Electromagnetic spectrum	The electromagnetic spectrum is the range of all possible frequencies of electromagnetic radiation. The 'electromagnetic spectrum' of an object is the characteristic distribution of electromagnetic radiation emitted or absorbed by that particular object. The electromagnetic spectrum extends from low frequencies used for modern radio communication to gamma radiation at the short-wavelength (high-frequency) end, thereby covering wavelengths from thousands of kilometers down to a fraction of the size of an atom.
Racemic mixture	In chemistry, a racemic mixture, is one that has equal amounts of left- and right-handed enantiomers of a chiral molecule. The first known racemic mixture was 'racemic acid', which Louis Pasteur found to be a mixture of the two enantiomeric isomers of tartaric acid. Nomenclature A racemic mixture is denoted by the prefix (±)- or dl- (for sugars the prefix DL- may be used), indicating an equal (1:1) mixture of dextro and levo isomers.

Chapter 2. Atoms: the foundations of life

Radio waves	Radio waves are a type of electromagnetic radiation with wavelengths in the electromagnetic spectrum longer than infrared light. Radio waves have frequencies from 300 GHz to as low as 3 kHz, and corresponding wavelengths from 1 millimeter to 100 kilometers. Like all other electromagnetic waves, they travel at the speed of light.
Wavelength	In physics, the wavelength of a sinusoidal wave is the spatial period of the wave--the distance over which the wave's shape repeats. It is usually determined by considering the distance between consecutive corresponding points of the same phase, such as crests, troughs, or zero crossings, and is a characteristic of both traveling waves and standing waves, as well as other spatial wave patterns. Wavelength is commonly designated by the Greek letter lambda (λ).
Lewis structure	Lewis structures (also known as Lewis dot diagrams, electron dot diagrams, and electron dot structures) are diagrams that show the bonding between atoms of a molecule and the lone pairs of electrons that may exist in the molecule.
Emission spectrum	The emission spectrum of a chemical element or chemical compound is the spectrum of frequencies of electromagnetic radiation emitted by the element's atoms or the compound's molecules when they are returned to a lower energy state. Each element's emission spectrum is unique. Therefore, spectroscopy can be used to identify the elements in matter of unknown composition.
Electron pair	In chemistry, an electron pair consists of two electrons that occupy the same orbital but have opposite spins. Because electrons are fermions, the Pauli exclusion principle forbids these particles from having exactly the same quantum numbers. Therefore the only way to occupy the same orbital, i.e. have the same orbital quantum numbers, is to differ in the spin quantum number.
Valence electron	In chemistry, valence electrons are the electrons of an atom that can participate in the formation of chemical bonds with other atoms. Valence electrons are their 'own' electrons, present in the free neutral atom, that combine with valence electrons of other atoms to form chemical bonds. In a single covalent bond both atoms contribute one valence electron to form a shared pair.

PRACTICE QUIZ
Chapter 2. Atoms: the foundations of life

1. In chemistry and physics, the _____ is the number of protons found in the nucleus of an atom and therefore identical to the charge number of the nucleus. It is conventionally represented by the symbol Z. The _____ uniquely identifies a chemical element. In an atom of neutral charge, the _____ is also equal to the number of electrons.

   a. Atomic weight
   b. Autoignition temperature
   c. Atomic number
   d. Electric torque

2. A _____ is a 1- or 2-letter internationally agreed code for a chemical element, usually derived from the name of the element, often in Latin.

   Only the first letter is capitalised. For example, 'He' is the symbol for helium, 'Pb' for lead, 'W' for tungsten.

   a. Systematic element name
   b. Barium
   c. Berkelium
   d. Chemical symbol

3. _____ is a process used by plants and other organisms to capture the sun's energy to split off water's hydrogen from oxygen. Hydrogen is combined with carbon dioxide (absorbed from air or water) to form glucose and release oxygen. All living cells in turn use fuels derived from glucose and oxidize the hydrogen and carbon to release the sun's energy and reform water and carbon dioxide in the process (cellular respiration).

   a. Plasmolysis
   b. Pressure Flow Hypothesis
   c. Photosynthesis
   d. Open shell

4. _____s for electrons in an atomic subshell are orbitals at identical energy levels. These are important in physical chemistry because they affect the ways electrons fill atoms. For example, all the 3p orbitals have same energy level, and so do all the 5d orbitals.

---

a. Kapustinskii equation
b. Degenerate orbital
c. Pseudorotation
d. Shared pair

5. In chemistry, a _____ (also known as a family) is a vertical column in the periodic table of the chemical elements. There are 18 groups in the standard periodic table, including the d-block elements, but excluding the f-block elements.

The explanation of the pattern of the table is that the elements in a group have similar physical or chemical characteristic of the outermost electron shells of their atoms (i.e. the same core charge), as most chemical properties are dominated by the orbital location of the outermost electron.

a. Boron group
b. Carbon group
c. Group
d. D-block

ANSWER KEY
Chapter 2. Atoms: the foundations of life

1. c
2. d
3. c
4. b
5. c

You can take the complete Chapter Practice Test

for Chapter 2. Atoms: the foundations of life
on all key terms, persons, places, and concepts.

Online 99 Cents

http://www.epub14.51.19910.2.cram101.com/

Use www.Cram101.com for all your study needs

including Cram101's online interactive problem solving labs in chemistry, statistics, mathematics, and more.

CHAPTER OUTLINE: KEY TERMS, PEOPLE, PLACES, CONCEPTS
Chapter 3
Compounds and chemical bonding: bringing atoms together

- _____ DNA profiling
- _____ Carbon dioxide
- _____ Chemical bond
- _____ Electron pair
- _____ Valence electron
- _____ Noble gas
- _____ Octet rule
- _____ Intramolecular force
- _____ Ionic compound
- _____ Covalent bond
- _____ Ionic bond
- _____ Ionization
- _____ Electron transfer
- _____ Chemical formula
- _____ Triple bond
- _____ Sodium chloride
- _____ Molecular orbital
- _____ Lewis structure
- _____ Atomic orbital

Chapter 3. Compounds and chemical bonding: bringing atoms together

- Helium
- Non-bonding orbital
- DNA-binding protein
- Pi bond
- Sigma bond
- X-ray crystallography
- Single bond
- Double bond
- Racemic mixture
- Carbon monoxide
- Benzene
- Conjugated system
- Ibuprofen
- Paracetamol
- Phenylalanine
- Calcium carbonate
- Polyatomic ion
- Electronegativity
- Bond cleavage

Chapter 3. Compounds and chemical bonding: bringing atoms together

Polar bond

CHAPTER HIGHLIGHTS: KEY TERMS, PEOPLE, PLACES, CONCEPTS
Chapter 3. Compounds and chemical bonding: bringing atoms together

DNA profiling	DNA profiling is a technique employed by forensic scientists to assist in the identification of individuals by their respective DNA profiles. DNA profiles are encrypted sets of numbers that reflect a person's DNA makeup, which can also be used as the person's identifier. DNA profiling should not be confused with full genome sequencing.
Carbon dioxide	Carbon dioxide is a naturally occurring chemical compound composed of two oxygen atoms covalently bonded to a single carbon atom. It is a gas at standard temperature and pressure and exists in Earth's atmosphere in this state, as a trace gas at a concentration of 0.039% by volume. As part of the carbon cycle known as photosynthesis, plants, algae, and cyanobacteria absorb carbon dioxide, light, and water to produce carbohydrate energy for themselves and oxygen as a waste product.
Chemical bond	A chemical bond is an attraction between atoms that allows the formation of chemical substances that contain two or more atoms. The bond is caused by the electromagnetic force attraction between opposite charges, either between electrons and nuclei, or as the result of a dipole attraction. The strength of chemical bonds varies considerably; there are 'strong bonds' such as covalent or ionic bonds and 'weak bonds' such as dipole-dipole interactions, the London dispersion force and hydrogen bonding.
Electron pair	In chemistry, an electron pair consists of two electrons that occupy the same orbital but have opposite spins. Because electrons are fermions, the Pauli exclusion principle forbids these particles from having exactly the same quantum numbers. Therefore the only way to occupy the same orbital, i.e. have the same orbital quantum numbers, is to differ in the spin quantum number.
Valence electron	In chemistry, valence electrons are the electrons of an atom that can participate in the formation of chemical bonds with other atoms. Valence electrons are their 'own' electrons, present in the free neutral atom, that combine with valence electrons of other atoms to form chemical bonds. In a single covalent bond both atoms contribute one valence electron to form a shared pair.

Chapter 3. Compounds and chemical bonding: bringing atoms together

Noble gas	The noble gases are a group of chemical elements with very similar properties: under standard conditions, they are all odorless, colorless, monatomic gases, with very low chemical reactivity. The six noble gases that occur naturally are helium (He), neon (Ne), argon (Ar), krypton (Kr), xenon (Xe), and the radioactive radon (Rn).
	For the first six periods of the periodic table, the noble gases are exactly the members of group 18 of the periodic table.
Octet rule	The octet rule is a chemical rule of thumb that states that atoms of low (<20) atomic number tend to combine in such a way that they each have eight electrons in their valence shells, giving them the same electronic configuration as a noble gas. The rule is applicable to the main-group elements, especially carbon, nitrogen, oxygen, and the halogens, but also to metals such as sodium or magnesium.
	The valence electrons can be counted using a Lewis electron dot diagram as shown at right for carbon dioxide.
Intramolecular force	An intramolecular force is any force that holds together the atoms making up a molecule or compound. They contain all types of chemical bond. They are stronger than intermolecular forces, which are present between atoms or molecules that are not actually bonded.
Ionic compound	In chemistry, an ionic compound is a chemical compound in which ions are held together in a lattice structure by ionic bonds. Usually, the positively charged portion consists of metal cations and the negatively charged portion is an anion or polyatomic ion. Ions in ionic compounds are held together by the electrostatic forces between oppositely charged bodies.
Covalent bond	A covalent bond is a form of chemical bonding that is characterized by the sharing of pairs of electrons between atoms. The stable balance of attractive and repulsive forces between atoms when they share electrons is known as covalent bonding.

Chapter 3. Compounds and chemical bonding: bringing atoms together

	Covalent bonding includes many kinds of interaction, including σ-bonding, π-bonding, metal-to-metal bonding, agostic interactions, and three-center two-electron bonds.
Ionic bond	An ionic bond is a type of chemical bond formed through an electrostatic attraction between two oppositely charged ions. Ionic bonds are formed between a cation, which is usually a metal, and an anion, which is usually a nonmetal. Pure ionic bonding cannot exist: all ionic compounds have some degree of covalent bonding.
Ionization	Ionization is the process of converting an atom or molecule into an ion by adding or removing charged particles such as electrons or ions. In the case of ionisation of a gas, ion-pairs are created consisting of a free electron and a +ve ion. Types of Ionisation The process of ionization works slightly differently depending on whether an ion with a positive or a negative electric charge is being produced.
Electron transfer	Electron transfer occurs when an electron moves from an atom or a chemical species (e.g. a molecule) to another atom or chemical species. ET is a mechanistic description of the thermodynamic concept of redox, wherein the oxidation states of both reaction partners change. Numerous biological processes involve ET reactions.
Chemical formula	A chemical formula is a way of expressing information about the atoms that constitute a particular chemical compound. The chemical formula identifies each constituent element by its chemical symbol and indicates the number of atoms of each element found in each discrete molecule of that compound. If a molecule contains more than one atom of a particular element, this quantity is indicated using a subscript after the chemical symbol (although 18th-century books often used superscripts) and also can be combined by more chemical elements.

Chapter 3. Compounds and chemical bonding: bringing atoms together

Triple bond	A triple bond in chemistry is a chemical bond between two chemical elements involving six bonding electrons instead of the usual two in a covalent single bond. The most common triple bond, that between two carbon atoms, can be found in alkynes. Other functional groups containing a triple bond are cyanides and isocyanides.
Sodium chloride	Sodium chloride, common salt, table salt or halite, is an ionic compound with the formula NaCl. Sodium chloride is the salt most responsible for the salinity of the ocean and of the extracellular fluid of many multicellular organisms. As the major ingredient in edible salt, it is commonly used as a condiment and food preservative.
Molecular orbital	In chemistry, a molecular orbital is a mathematical function describing the wave-like behavior of an electron in a molecule. This function can be used to calculate chemical and physical properties such as the probability of finding an electron in any specific region. The term 'orbital' was first used in English by Robert S. Mulliken as the English translation of Schrödinger's 'Eigenfunktion'.
Lewis structure	Lewis structures (also known as Lewis dot diagrams, electron dot diagrams, and electron dot structures) are diagrams that show the bonding between atoms of a molecule and the lone pairs of electrons that may exist in the molecule.
Atomic orbital	An atomic orbital is a mathematical function that describes the wave-like behavior of either one electron or a pair of electrons in an atom. This function can be used to calculate the probability of finding any electron of an atom in any specific region around the atom's nucleus. The term may also refer to the physical region where the electron can be calculated to be, as defined by the particular mathematical form of the orbital.
Helium	Helium is the chemical element with atomic number 2 and an atomic weight of 4.002602, which is represented by the symbol He. It is a colorless, odorless, tasteless, non-toxic, inert, monatomic gas that heads the noble gas group in the periodic table. Its boiling and melting points are the lowest among the elements and it exists only as a gas except in extreme conditions.
Non-bonding orbital	A non-bonding orbital, is a molecular orbital whose occupation by electrons neither increases nor decreases the bond order between the involved atoms. Non-bonding orbitals are often designated by the letter n in molecular orbital diagrams and electron transition notations. Non-bonding orbitals are the equivalent in molecular orbital theory of the lone pairs in Lewis structures.

Chapter 3. Compounds and chemical bonding: bringing atoms together

DNA-binding protein	DNA-binding proteins are proteins that are composed of DNA-binding domains and thus have a specific or general affinity for either single or double stranded DNA. Sequence-specific DNA-binding proteins generally interact with the major groove of B-DNA, because it exposes more functional groups that identify a base pair. However there are some known minor groove DNA-binding ligands such as Netropsin, Distamycin, Hoechst 33258, Pentamidine and others. Examples DNA-binding proteins include transcription factors which modulate the process of transcription, various polymerases, nucleases which cleave DNA molecules, and histones which are involved in chromosome packaging and transcription in the cell nucleus.
Pi bond	In chemistry, pi bonds (π bonds) are covalent chemical bonds where two lobes of one involved atomic orbital overlap two lobes of the other involved atomic orbital. These orbitals share a nodal plane which passes through both of the involved nuclei. The Greek letter π in their name refers to p orbitals, since the orbital symmetry of the pi bond is the same as that of the p orbital when seen down the bond axis.
Sigma bond	In chemistry, sigma bonds (σ bonds) are the strongest type of covalent chemical bond. They are formed by head-on overlapping between atomic orbitals. Sigma bonding is most clearly defined for diatomic molecules using the language and tools of symmetry groups.
X-ray crystallography	X-ray crystallography is a method of determining the arrangement of atoms within a crystal, in which a beam of X-rays strikes a crystal and causes the beam of light to spread into many specific directions. From the angles and intensities of these diffracted beams, a crystallographer can produce a three-dimensional picture of the density of electrons within the crystal. From this electron density, the mean positions of the atoms in the crystal can be determined, as well as their chemical bonds, their disorder and various other information.
Single bond	A Single bond in chemistry is a chemical bond between two chemical elements involving two bonding electrons.

Chapter 3. Compounds and chemical bonding: bringing atoms together

	Usually, Single bond is Sigma bond. but diboron is Pi bond.(Molecular orbital diagram#Diboron MO diagram.
Double bond	A double bond in chemistry is a chemical bond between two chemical elements involving four bonding electrons instead of the usual two. The most common double bond, that between two carbon atoms, can be found in alkenes. Many types of double bonds between two different elements exist, for example in a carbonyl group with a carbon atom and an oxygen atom.
Racemic mixture	In chemistry, a racemic mixture, is one that has equal amounts of left- and right-handed enantiomers of a chiral molecule. The first known racemic mixture was 'racemic acid', which Louis Pasteur found to be a mixture of the two enantiomeric isomers of tartaric acid. Nomenclature A racemic mixture is denoted by the prefix (±)- or dl- (for sugars the prefix DL- may be used), indicating an equal (1:1) mixture of dextro and levo isomers.
Carbon monoxide	Carbon monoxide also called carbonous oxide, is a colorless, odorless, and tasteless gas which is slightly lighter than air. It is highly toxic to humans and animals in higher quantities, although it is also produced in normal animal metabolism in low quantities, and is thought to have some normal biological functions. Carbon monoxide consists of one carbon atom and one oxygen atom, connected by a triple bond which consists of two covalent bonds as well as one dative covalent bond.
Benzene	Benzene is an organic chemical compound with the molecular formula C_6H_6. Its molecule is composed of 6 carbon atoms joined in a ring, with 1 hydrogen atom attached to each carbon atom. Because its molecules contain only carbon and hydrogen atoms, benzene is classed as a hydrocarbon.
Conjugated system	In chemistry, a conjugated system is a system of connected p-orbitals with delocalized electrons in compounds with alternating single and multiple bonds, which in general may lower the overall energy of the molecule and increase stability. Lone pairs, radicals or carbenium ions may be part of the system. The compound may be cyclic, acyclic, linear or mixed.

Chapter 3. Compounds and chemical bonding: bringing atoms together

Ibuprofen	Ibuprofen ('a?ju?pro?f?n or a?ju?'pro?f?n-bew-PROH-f?n; from the nomenclature iso-butyl-propanoic-phenolic acid) is a nonsteroidal anti-inflammatory drug (NSAID) used for relief of symptoms of arthritis, fever, as an analgesic (pain reliever), especially where there is an inflammatory component, and dysmenorrhea.
	Ibuprofen is known to have an antiplatelet effect, though it is relatively mild and somewhat short-lived when compared with aspirin or other better-known antiplatelet drugs. In general, ibuprofen also acts as a vasoconstrictor, having been shown to constrict coronary arteries and some other blood vessels mainly because it inhibits the vasodilating prostacyclin produced by cyclooxygenase 2 enzymes.
Paracetamol	Paracetamol INN (?pær?'si?t?m?l or ?pær?'s?t?m?l), or acetaminophen USAN ??si?t?'m??f?n, is a widely used over-the-counter analgesic (pain reliever) and antipyretic (fever reducer). It is commonly used for the relief of headaches and other minor aches and pains and is a major ingredient in numerous cold and flu remedies. In combination with opioid analgesics, paracetamol can also be used in the management of more severe pain such as post-surgical pain and providing palliative care in advanced cancer patients.
Phenylalanine	Phenylalanine is an α-amino acid with the formula $HO_2CCH(NH_2)CH_2C_6H_5$. This essential amino acid is classified as nonpolar because of the hydrophobic nature of the benzyl side chain. L-Phenylalanine is an electrically neutral amino acid, one of the twenty common amino acids used to biochemically form proteins, coded for by DNA. The codons for L-phenylalanine are UUU and UUC. Phenylalanine is a precursor for tyrosine, the monoamine signaling molecules dopamine, norepinephrine (noradrenaline), and epinephrine (adrenaline), and the skin pigment melanin.
Calcium carbonate	Calcium carbonate is a chemical compound with the formula $CaCO_3$. It is a common substance found in rocks in all parts of the world, and is the main component of shells of marine organisms, snails, coal balls, pearls, and eggshells. Calcium carbonate is the active ingredient in agricultural lime, and is usually the principal cause of hard water.
Polyatomic ion	A polyatomic ion, is a charged species (ion) composed of two or more atoms covalently bonded or of a metal complex that can be considered as acting as a single unit in the context of acid and base chemistry or in the formation of salts. The prefix 'poly-' means 'many,' in Greek, but even ions of two atoms are commonly referred to as polyatomic. In older literature, a polyatomic ion is also referred to as a radical, and less commonly, as a radical group.

Chapter 3. Compounds and chemical bonding: bringing atoms together

Electronegativity	Electronegativity, symbol χ, is a chemical property that describes the tendency of an atom or a functional group to attract electrons towards itself. An atom's electronegativity is affected by both its atomic number and the distance that its valence electrons reside from the charged nucleus. The higher the associated electronegativity number, the more an element or compound attracts electrons towards it.
Bond cleavage	Bond cleavage, is the splitting of chemical bonds. If the two electrons in a cleaved covalent bond are divided between the products, the process is known as homolytic fission or homolysis and free redicals are generated by homolytic cleavage. Alternatively, the case where both electrons are retained by one product and charged species that is nucleophile and electrophile are generated by the process is known as heterolytic fission and (heterolysis).
Polar bond	In chemistry, a polar bond is a type of covalent bond between two atoms or more in which electrons are shared unequally. Because of this, one end of the molecule has a slight, relative negative charge and the other a slight, relative positive charge. An example of atoms bonded by a polar bond is the water molecule, which is made up of two hydrogen atoms and one oxygen atom.

PRACTICE QUIZ
Chapter 3. Compounds and chemical bonding: bringing atoms together

1. In chemistry, a _____, is one that has equal amounts of left- and right-handed enantiomers of a chiral molecule. The first known _____ was 'racemic acid', which Louis Pasteur found to be a mixture of the two enantiomeric isomers of tartaric acid.

    Nomenclature
    A _____ is denoted by the prefix (±)- or dl- (for sugars the prefix DL- may be used), indicating an equal (1:1) mixture of dextro and levo isomers.

    a. Racemization
    b. Ray-Dutt twist
    c. Racemic mixture
    d. Ring flip

2. _____ also called carbonous oxide, is a colorless, odorless, and tasteless gas which is slightly lighter than air. It is highly toxic to humans and animals in higher quantities, although it is also produced in normal animal metabolism in low quantities, and is thought to have some normal biological functions.

    _____ consists of one carbon atom and one oxygen atom, connected by a triple bond which consists of two covalent bonds as well as one dative covalent bond.

    a. Herbicide
    b. Gentamicin
    c. Carbon monoxide
    d. Ring flip

3. _____ is the chemical element with atomic number 2 and an atomic weight of 4.002602, which is represented by the symbol He. It is a colorless, odorless, tasteless, non-toxic, inert, monatomic gas that heads the noble gas group in the periodic table. Its boiling and melting points are the lowest among the elements and it exists only as a gas except in extreme conditions.

---

a. Liquid helium
   b. Krypton
   c. Helium
   d. Penning mixture

4. In chemistry, an _____ is a chemical compound in which ions are held together in a lattice structure by ionic bonds. Usually, the positively charged portion consists of metal cations and the negatively charged portion is an anion or polyatomic ion. Ions in _____s are held together by the electrostatic forces between oppositely charged bodies.

   a. Ionic liquid
   b. Ionic potential
   c. Ionic transfer
   d. Ionic compound

5. The _____es are a group of chemical elements with very similar properties: under standard conditions, they are all odorless, colorless, monatomic gases, with very low chemical reactivity. The six _____es that occur naturally are helium (He), neon (Ne), argon (Ar), krypton (Kr), xenon (Xe), and the radioactive radon (Rn).

   For the first six periods of the periodic table, the _____es are exactly the members of group 18 of the periodic table.

   a. 1,2-Dioxetanedione
   b. Van der Waals force
   c. Noble gas
   d. Walsh diagram

ANSWER KEY
Chapter 3. Compounds and chemical bonding: bringing atoms together

1. c
2. c
3. c
4. d
5. c

You can take the complete Chapter Practice Test

for Chapter 3. Compounds and chemical bonding: bringing atoms together
on all key terms, persons, places, and concepts.

Online 99 Cents

http://www.epub14.51.19910.3.cram101.com/

Use www.Cram101.com for all your study needs

including Cram101's online interactive problem solving labs in chemistry, statistics, mathematics, and more.

CHAPTER OUTLINE: KEY TERMS, PEOPLE, PLACES, CONCEPTS
Chapter 4
Molecular interactions: holding it all together

- _____ DNA profiling
- _____ Intramolecular force
- _____ Intermolecular force
- _____ Endocrine system
- _____ Hormone
- _____ ATP synthase
- _____ Polar bond
- _____ Racemic mixture
- _____ Molecular geometry
- _____ Permanent
- _____ X-ray crystallography
- _____ Diatomic molecule
- _____ Geometry
- _____ Avogadro constant
- _____ Cytosine
- _____ Double Helix
- _____ Guanine
- _____ Aqueous solution
- _____ Ionic bond

Chapter 4. Molecular interactions: holding it all together

- _____ Ionization
- _____ Salt bridge
- _____ Solvation
- _____ Solvent
- _____ Hydrophobicity
- _____ Solubility
- _____ Thermodynamics
- _____ Cell membrane
- _____ Lipid bilayer
- _____ Alcohol
- _____ Melting point
- _____ Boiling point
- _____ Condensation
- _____ Melting
- _____ Vaporization

CHAPTER HIGHLIGHTS: KEY TERMS, PEOPLE, PLACES, CONCEPTS
Chapter 4. Molecular interactions: holding it all together

DNA profiling	DNA profiling is a technique employed by forensic scientists to assist in the identification of individuals by their respective DNA profiles. DNA profiles are encrypted sets of numbers that reflect a person's DNA makeup, which can also be used as the person's identifier. DNA profiling should not be confused with full genome sequencing.
Intramolecular force	An intramolecular force is any force that holds together the atoms making up a molecule or compound. They contain all types of chemical bond. They are stronger than intermolecular forces, which are present between atoms or molecules that are not actually bonded.
Intermolecular force	Intermolecular forces are forces of attraction or repulsion which act between neighboring particles: atoms, molecules or ions. They are weak compared to the intramolecular forces, the forces which keep a molecule together. For example, the covalent bond present within HCl molecules is much stronger than the forces present between the neighbouring molecules, which exist when the molecules are sufficiently close to each other.
Endocrine system	The endocrine system is the system of glands, each of which secretes a type of hormone directly into the bloodstream to regulate the body. The endocrine system is in contrast to the exocrine system, which secretes its chemicals using ducts. It derives from the Greek words 'endo' meaning inside, within, and 'crinis' for secrete.
Hormone	A hormone is a chemical released by a cell or a gland in one part of the body that sends out messages that affect cells in other parts of the organism. Only a small amount of hormone is required to alter cell metabolism. In essence, it is a chemical messenger that transports a signal from one cell to another.
ATP synthase	ATP synthase is an important enzyme that provides energy for the cell to use through the synthesis of adenosine triphosphate (ATP). ATP is the most commonly used 'energy currency' of cells from most organisms. It is formed from adenosine diphosphate (ADP) and inorganic phosphate (P_i), and needs energy.
Polar bond	In chemistry, a polar bond is a type of covalent bond between two atoms or more in which electrons are shared unequally. Because of this, one end of the molecule has a slight, relative negative charge and the other a slight, relative positive charge. An example of atoms bonded by a polar bond is the water molecule, which is made up of two hydrogen atoms and one oxygen atom.
Racemic mixture	In chemistry, a racemic mixture, is one that has equal amounts of left- and right-handed enantiomers of a chiral molecule. The first known racemic mixture was 'racemic acid', which Louis Pasteur found to be a mixture of the two enantiomeric isomers of tartaric acid.
	Nomenclature

Chapter 4. Molecular interactions: holding it all together

	A racemic mixture is denoted by the prefix (±)- or dl- (for sugars the prefix DL- may be used), indicating an equal (1:1) mixture of dextro and levo isomers.
Molecular geometry	Molecular geometry is the three-dimensional arrangement of the atoms that constitute a molecule. It determines several properties of a substance including its reactivity, polarity, phase of matter, color, magnetism, and biological activity. The molecular geometry can be determined by various spectroscopic methods and diffraction methods.
Permanent	The permanent of a square matrix in linear algebra is a function of the matrix similar to the determinant. The permanent, as well as the determinant, is a polynomial in the entries of the matrix. Both permanent and determinant are special cases of a more general function of a matrix called the immanant.
X-ray crystallography	X-ray crystallography is a method of determining the arrangement of atoms within a crystal, in which a beam of X-rays strikes a crystal and causes the beam of light to spread into many specific directions. From the angles and intensities of these diffracted beams, a crystallographer can produce a three-dimensional picture of the density of electrons within the crystal. From this electron density, the mean positions of the atoms in the crystal can be determined, as well as their chemical bonds, their disorder and various other information.
Diatomic molecule	Diatomic molecules are molecules composed only of two atoms, of either the same or different chemical elements. The prefix di- is of Greek origin, meaning 2. Common diatomic molecules are hydrogen (H_2), nitrogen (N_2), oxygen (O_2), and carbon monoxide (CO). Seven elements exist as homonuclear diatomic molecules at room temperature: H_2, N_2, O_2, F_2, Cl_2, Br_2, and I_2.
Geometry	Geometry is a branch of mathematics concerned with questions of shape, size, relative position of figures, and the properties of space. A mathematician who works in the field of geometry is called a geometer. Geometry arose independently in a number of early cultures as a body of practical knowledge concerning lengths, areas, and volumes, with elements of a formal mathematical science emerging in the West as early as Thales (6th Century BC).

Chapter 4. Molecular interactions: holding it all together

Avogadro constant	In chemistry and physics, the Avogadro constant is defined as the ratio of the number of constituent particles N (usually atoms or molecules) in a sample to the amount of substance n (unit mole) through the relationship $N_A = Nn$. Thus, it is the proportionality factor that relates the molar mass of an entity, i.e., the mass per amount of substance, to the mass of said entity. The Avogadro constant expresses the number of elementary entities per mole of substance and it has the value 6.022 141 29(27) × 10^{23} mol^{-1}.
Cytosine	Cytosine is one of the four main bases found in DNA and RNA, along with adenine, guanine, and thymine (uracil in RNA). It is a pyrimidine derivative, with a heterocyclic aromatic ring and two substituents attached (an amine group at position 4 and a keto group at position 2). The nucleoside of cytosine is cytidine.
Double Helix	Double Helix (2004), a novel by Nancy Werlin, is about 18-year old Eli Samuels, who works for a famous molecular biologist named Dr. Quincy Wyatt. There is a mysterious connection between Dr. Wyatt and Eli's parents, and all Eli knows about the connection is that it has something to do with his mother, who has Huntington's disease. Because of the connection between Dr. Wyatt and the Samuels family, Eli's father is strongly against Eli working there.
Guanine	Guanine is one of the four main nucleobases found in the nucleic acids DNA and RNA, the others being adenine, cytosine, and thymine (uracil in RNA). In DNA, guanine is paired with cytosine. With the formula $C_5H_5N_5O$, guanine is a derivative of purine, consisting of a fused pyrimidine-imidazole ring system with conjugated double bonds.
Aqueous solution	An aqueous solution is a solution in which the solvent is water. It is usually shown in chemical equations by appending (aq) to the relevant formula, such as NaCl (aq). The word aqueous means pertaining to, related to, similar to, or dissolved in water.
Ionic bond	An ionic bond is a type of chemical bond formed through an electrostatic attraction between two oppositely charged ions. Ionic bonds are formed between a cation, which is usually a metal, and an anion, which is usually a nonmetal. Pure ionic bonding cannot exist: all ionic compounds have some degree of covalent bonding.
Ionization	Ionization is the process of converting an atom or molecule into an ion by adding or removing charged particles such as electrons or ions. In the case of ionisation of a gas, ion-pairs are created consisting of a free electron and a +ve ion. Types of Ionisation

Chapter 4. Molecular interactions: holding it all together

	The process of ionization works slightly differently depending on whether an ion with a positive or a negative electric charge is being produced.
Salt bridge	A salt bridge, in chemistry, is a laboratory device used to connect the oxidation and reduction half-cells of a galvanic cell (voltaic cell), a type of electrochemical cell. Salt bridges usually come in two types: glass tube and filter paper. Glass tube bridges One type of salt bridge consists of a U-shaped glass tube filled with a relatively inert electrolyte; usually potassium chloride or sodium chloride is used, although the diagram here illustrates the use of a potassium nitrate solution.
Solvation	Solvation, also sometimes called dissolution, is the process of attraction and association of molecules of a solvent with molecules or ions of a solute. As ions dissolve in a solvent they spread out and become surrounded by solvent molecules. Distinction between solvation, dissolution and solubility By an IUPAC definition, solvation is an interaction of a solute with the solvent, which leads to stabilization of the solute species in the solution.
Solvent	A solvent is a liquid, solid, or gas that dissolves another solid, liquid, or gaseous solute, resulting in a solution that is soluble in a certain volume of solvent at a specified temperature. Common uses for organic solvents are in dry cleaning (e.g., tetrachloroethylene), as paint thinners (e.g., toluene, turpentine), as nail polish removers and glue solvents (acetone, methyl acetate, ethyl acetate), in spot removers (e.g., hexane, petrol ether), in detergents (citrus terpenes), in perfumes (ethanol), nail polish and in chemical synthesis. The use of inorganic solvents (other than water) is typically limited to research chemistry and some technological processes.
Hydrophobicity	In chemistry, hydrophobicity is the physical property of a molecule (known as a hydrophobe) that is repelled from a mass of water.

Chapter 4. Molecular interactions: holding it all together

	Hydrophobic molecules tend to be non-polar and thus prefer other neutral molecules and non-polar solvents. Hydrophobic molecules in water often cluster together forming micelles.
Solubility	Solubility is the property of a solid, liquid, or gaseous chemical substance called solute to dissolve in a solid, liquid, or gaseous solvent to form a homogeneous solution of the solute in the solvent. The solubility of a substance fundamentally depends on the used solvent as well as on temperature and pressure. The extent of the solubility of a substance in a specific solvent is measured as the saturation concentration, where adding more solute does not increase the concentration of the solution.
Thermodynamics	Property database Thermodynamics is the branch of physical science concerned with heat and its relation to other forms of energy and work. It defines macroscopic variables (such as temperature, entropy, and pressure) that describe average properties of material bodies and radiation, and explains how they are related and by what laws they change with time. Thermodynamics does not describe the microscopic constituents of matter, and its laws can be derived from statistical mechanics.
Cell membrane	The cell membrane is a biological membrane that separates the interior of all cells from the outside environment. The cell membrane is selectively permeable to ions and organic molecules and controls the movement of substances in and out of cells. It basically protects the cell from outside forces.
Lipid bilayer	The lipid bilayer is a thin polar membrane made of two layers of lipid molecules. These membranes are flat sheets that form a continuous barrier around cells. The cell membrane of almost all living organisms and many viruses are made of a lipid bilayer, as are the membranes surrounding the cell nucleus and other sub-cellular structures.
Alcohol	In chemistry, an alcohol is an organic compound in which the hydroxyl functional group (-OH) is bound to a carbon atom. In particular, this carbon center should be saturated, having single bonds to three other atoms.

Chapter 4. Molecular interactions: holding it all together

An important class of alcohols are the simple acyclic alcohols, the general formula for which is $C_nH_{2n+1}OH$. Of those, ethanol (C_2H_5OH) is the type of alcohol found in alcoholic beverages, and in common speech the word alcohol refers specifically to ethanol.

Melting point — The melting point of a solid is the temperature at which it changes state from solid to liquid. At the melting point the solid and liquid phase exist in equilibrium. The melting point of a substance depends (usually slightly) on pressure and is usually specified at standard pressure.

Boiling point — The boiling point of a substance is the temperature at which the vapor pressure of the liquid equals the environmental pressure surrounding the liquid.

A liquid in a vacuum has a lower boiling point than when that liquid is at atmospheric pressure. A liquid at high-pressure has a higher boiling point than when that liquid is at atmospheric pressure.

Condensation — Condensation is the change of the physical state of matter from gaseous phase into liquid phase, and is the reverse of vaporization. When the transition happens from the gaseous phase into the solid phase directly, the change is called deposition.

Condensation is initiated by the formation of atomic/molecular clusters of that species within its gaseous volume--like rain drop or snow-flake formation within clouds--or at the contact between such gaseous phase and a (solvent) liquid or solid surface.

Melting — Melting, is a physical process that results in the phase transition of a substance from a solid to a liquid. The internal energy of a substance is increased, typically by the application of heat or pressure, resulting in a rise of its temperature to the melting point, at which the rigid ordering of molecular entities in the solid breaks down to a less-ordered state and the solid liquefies. An object that has melted completely is molten.

Chapter 4. Molecular interactions: holding it all together

Vaporization	Vaporization of an element or compound is a phase transition from the liquid phase to gas phase. There are two types of vaporization: evaporation and boiling.
	Evaporation is a phase transition from the liquid phase to gas phase that occurs at temperatures below the boiling temperature at a given pressure.

PRACTICE QUIZ
Chapter 4. Molecular interactions: holding it all together

1. In chemistry, _____ is the physical property of a molecule (known as a hydrophobe) that is repelled from a mass of water.

   Hydrophobic molecules tend to be non-polar and thus prefer other neutral molecules and non-polar solvents. Hydrophobic molecules in water often cluster together forming micelles.

   a. Hydrophobicity
   b. Van der Waals force
   c. Hydrogen bond
   d. Buckingham potential

2. _____ is a technique employed by forensic scientists to assist in the identification of individuals by their respective DNA profiles. DNA profiles are encrypted sets of numbers that reflect a person's DNA makeup, which can also be used as the person's identifier. _____ should not be confused with full genome sequencing.

   a. DNA sequencer
   b. DNA profiling
   c. DNA supercoil
   d. DNA-DNA hybridization

3. _____s are forces of attraction or repulsion which act between neighboring particles: atoms, molecules or ions. They are weak compared to the intramolecular forces, the forces which keep a molecule together. For example, the covalent bond present within HCl molecules is much stronger than the forces present between the neighbouring molecules, which exist when the molecules are sufficiently close to each other.

   a. Intermolecular force
   b. Intramolecular force
   c. Ionic bond
   d. Isopeptide bond

4. In chemistry, an _____ is an organic compound in which the hydroxyl functional group (-OH) is bound to a carbon atom. In particular, this carbon center should be saturated, having single bonds to three other atoms.

An important class of _____s are the simple acyclic _____s, the general formula for which is $C_nH_{2n+1}OH$. Of those, ethanol (C_2H_5OH) is the type of _____ found in alcoholic beverages, and in common speech the word _____ refers specifically to ethanol.

a. Aldehyde
b. Aldimine
c. Alkane
d. Alcohol

5. _____ is the three-dimensional arrangement of the atoms that constitute a molecule. It determines several properties of a substance including its reactivity, polarity, phase of matter, color, magnetism, and biological activity.

The _____ can be determined by various spectroscopic methods and diffraction methods.

a. Molecular geometry
b. Bartell mechanism
c. Berry mechanism
d. Bond length

ANSWER KEY
Chapter 4. Molecular interactions: holding it all together

1. a
2. b
3. a
4. d
5. a

You can take the complete Chapter Practice Test

for Chapter 4. Molecular interactions: holding it all together

on all key terms, persons, places, and concepts.

Online 99 Cents

http://www.epub14.51.19910.4.cram101.com/

Use www.Cram101.com for all your study needs

including Cram101's online interactive problem solving labs in chemistry, statistics, mathematics, and more.

CHAPTER OUTLINE: KEY TERMS, PEOPLE, PLACES, CONCEPTS
Chapter 5
Organic compounds 1: the framework of life

- _____ Organic chemistry
- _____ Organic compound
- _____ Vitamin C
- _____ DNA profiling
- _____ Carbon
- _____ Carbon cycle
- _____ Elemental
- _____ Environmental Chemistry
- _____ Fullerene
- _____ Graphite
- _____ X-ray crystallography
- _____ Fossil fuel
- _____ Octet rule
- _____ Functional group
- _____ Alkane
- _____ Alkene
- _____ Alkyne
- _____ Hydrocarbon
- _____ Structural formula

Chapter 5. Organic compounds 1: the framework of life

- _____ Greenhouse gas
- _____ Anabolic steroid
- _____ Boiling point
- _____ Melting point
- _____ Progesterone
- _____ Steroid
- _____ Solubility
- _____ DNA-binding protein
- _____ Combustion
- _____ Intramolecular force
- _____ Double bond
- _____ Triple bond
- _____ Lewis structure
- _____ Genetic engineering
- _____ Geraniol
- _____ Hormone
- _____ Limonene
- _____ Epigenetics
- _____ Gene expression

Chapter 5. Organic compounds 1: the framework of life

_____ | Histone

_____ | Methyl group

_____ | Acetic acid

_____ | Benzene

_____ | Phenyl group

_____ | Electronegativity

_____ | Cell membrane

CHAPTER HIGHLIGHTS: KEY TERMS, PEOPLE, PLACES, CONCEPTS
Chapter 5. Organic compounds 1: the framework of life

Organic chemistry	Organic chemistry is a subdiscipline within chemistry involving the scientific study of the structure, properties, composition, reactions, and preparation (by synthesis or by other means) of carbon-based compounds, hydrocarbons, and their derivatives. These compounds may contain any number of other elements, including hydrogen, nitrogen, oxygen, the halogens as well as phosphorus, silicon, and sulfur. Organic compounds are structurally diverse.
Organic compound	An organic compound is any member of a large class of gaseous, liquid, or solid chemical compounds whose molecules contain carbon. For historical reasons discussed below, a few types of carbon-containing compounds such as carbides, carbonates, simple oxides of carbon, and cyanides, as well as the allotropes of carbon such as diamond and graphite, are considered inorganic. The distinction between 'organic' and 'inorganic' carbon compounds, while 'useful in organizing the vast subject of chemistry... is somewhat arbitrary'.
Vitamin C	Vitamin C is an essential nutrient for humans and certain other animal species. In living organisms ascorbate acts as an antioxidant by protecting the body against oxidative stress. It is also a cofactor in at least eight enzymatic reactions including several collagen synthesis reactions that, when dysfunctional, cause the most severe symptoms of scurvy.
DNA profiling	DNA profiling is a technique employed by forensic scientists to assist in the identification of individuals by their respective DNA profiles. DNA profiles are encrypted sets of numbers that reflect a person's DNA makeup, which can also be used as the person's identifier. DNA profiling should not be confused with full genome sequencing.
Carbon	Carbon 'k?rb?n is the chemical element with symbol C and atomic number 6. As a member of group 14 on the periodic table, it is nonmetallic and tetravalent--making four electrons available to form covalent chemical bonds. There are three naturally occurring isotopes, with ^{12}C and ^{13}C being stable, while ^{14}C is radioactive, decaying with a half-life of about 5,730 years. Carbon is one of the few elements known since antiquity.

Chapter 5. Organic compounds 1: the framework of life

Carbon cycle	The carbon cycle is the biogeochemical cycle by which carbon is exchanged among the biosphere, pedosphere, geosphere, hydrosphere, and atmosphere of the Earth. It is one of the most important cycles of the Earth and allows for carbon to be recycled and reused throughout the biosphere and all of its organisms. The global carbon budget is the balance of the exchanges (incomes and losses) of carbon between the carbon reservoirs or between one specific loop (e.g., atmosphere ↔ biosphere) of the carbon cycle.
Elemental	An Elemental is a spirit embodying one of the five elements of antiquity: Earth (solid), Water (liquid), Wind (gas), Fire (heat), Aether (quintessence). Elementals are referred to by various names. In the English European tradition these include Fairies, Elves, Devas, Brownies, Leprechauns, Gnomes, Sprites, Pixies, Banshees, Goblins, Dryads, Mermaids, Trolls, Dragons, Griffins, and numerous others.
Environmental Chemistry	Environmental Chemistry - Environmental problems, chemical approaches - is a peer-reviewed scientific journal publishing original research and review articles on all aspects of environmental chemistry. The Journal publishes manuscripts addressing the chemistry of the environment (air, water, soil, sediments, space, and biota) including atmospheric chemistry, (bio)geochemistry, climate change, marine chemistry, water chemistry, polar chemistry, fire chemistry, astrochemistry, earth and geochemistry, soil and sediment chemistry and chemical toxicology. Papers are written in a style that is accessible to those outside the field, as the readership will include chemists, biologists, toxicologists, soil scientists, and workers from government and industrial institutions.
Fullerene	The smallest fullerene is the dodecahedral C_{20}. There are no fullerenes with 22 vertices. The number of fullerenes C_{2n} grows with increasing n = 12, 13, 14, .. roughly in proportion to n^9 (sequence A007894 in OEIS).

Chapter 5. Organic compounds 1: the framework of life

Graphite	The mineral graphite is an allotrope of carbon. It was named by Abraham Gottlob Werner in 1789 from the Ancient Greek γρ?φω (grapho), 'to draw/write', for its use in pencils, where it is commonly called lead (not to be confused with the metallic element lead). Unlike diamond (another carbon allotrope), graphite is an electrical conductor, a semimetal.
X-ray crystallography	X-ray crystallography is a method of determining the arrangement of atoms within a crystal, in which a beam of X-rays strikes a crystal and causes the beam of light to spread into many specific directions. From the angles and intensities of these diffracted beams, a crystallographer can produce a three-dimensional picture of the density of electrons within the crystal. From this electron density, the mean positions of the atoms in the crystal can be determined, as well as their chemical bonds, their disorder and various other information.
Fossil fuel	Fossil fuels are fuels formed by natural resources such as anaerobic decomposition of buried dead organisms. The age of the organisms and their resulting fossil fuels is typically millions of years, and sometimes exceeds 650 million years. The fossil fuels, which contain high percentages of carbon, include coal, petroleum, and natural gas.
Octet rule	The octet rule is a chemical rule of thumb that states that atoms of low (<20) atomic number tend to combine in such a way that they each have eight electrons in their valence shells, giving them the same electronic configuration as a noble gas. The rule is applicable to the main-group elements, especially carbon, nitrogen, oxygen, and the halogens, but also to metals such as sodium or magnesium. The valence electrons can be counted using a Lewis electron dot diagram as shown at right for carbon dioxide.
Functional group	In organic chemistry, functional groups are lexicon specific groups of atoms or bonds within molecules that are responsible for the characteristic chemical reactions of those molecules. The same functional group will undergo the same or similar chemical reaction(s) regardless of the size of the molecule it is a part of. However, its relative reactivity can be modified by nearby functional groups.

Chapter 5. Organic compounds 1: the framework of life

Alkane	Alkanes (also known as paraffins or saturated hydrocarbons) are chemical compounds that consist only of hydrogen and carbon atoms and are bonded exclusively by single bonds (i.e., they are saturated compounds) without any cycles . Alkanes belong to a homologous series of organic compounds in which the members differ by a constant relative molecular mass of 14. They have 2 main commercial sources, crude oil and natural gas.
	Each carbon atom has 4 bonds (either C-H or C-C bonds), and each hydrogen atom is joined to a carbon atom (H-C bonds).
Alkene	In organic chemistry, an alkene, olefin, or olefine is an unsaturated chemical compound containing at least one carbon-to-carbon double bond. The simplest acyclic alkenes, with only one double bond and no other functional groups, form an homologous series of hydrocarbons with the general formula C_nH_{2n}.
	The simplest alkene is ethylene (C_2H_4), which has the International Union of Pure and Applied Chemistry (IUPAC) name ethene.
Alkyne	Alkynes are hydrocarbons that have a triple bond between two carbon atoms, with the formula C_nH_{2n-2}. Alkynes are traditionally known as acetylenes, although the name acetylene also refers specifically to C_2H_2, known formally as ethyne using IUPAC nomenclature. Like other hydrocarbons, alkynes are generally hydrophobic but tend to be more reactive.
Hydrocarbon	In organic chemistry, a hydrocarbon is an organic compound consisting entirely of hydrogen and carbon. Hydrocarbons from which one hydrogen atom has been removed are functional groups, called hydrocarbyls. Aromatic hydrocarbons (arenes), alkanes, alkenes, cycloalkanes and alkyne-based compounds are different types of hydrocarbons.
Structural formula	The structural formula of a chemical compound is a graphical representation of the molecular structure, showing how the atoms are arranged. The chemical bonding within the molecule is also shown, either explicitly or implicitly. Also several other formats are used, as in chemical databases, such as SMILES, InChI and CML.

Chapter 5. Organic compounds 1: the framework of life

	Unlike chemical formulas or chemical names, structural formulas provide a representation of the molecular structure.
Greenhouse gas	A greenhouse gas is a gas in an atmosphere that absorbs and emits radiation within the thermal infrared range. This process is the fundamental cause of the greenhouse effect. The primary greenhouse gases in the Earth's atmosphere are water vapour, carbon dioxide, methane, nitrous oxide, and ozone.
Anabolic steroid	Anabolic steroids, technically known as anabolic-androgen steroids (AAS) or colloquially as 'steroids', are drugs that mimic the effects of testosterone and dihydrotestosterone in the body. They increase protein synthesis within cells, which results in the buildup of cellular tissue (anabolism), especially in muscles. Anabolic steroids also have androgenic and virilizing properties, including the development and maintenance of masculine characteristics such as the growth of the vocal cords, testicles, and body hair (secondary sexual characteristics).
Boiling point	The boiling point of a substance is the temperature at which the vapor pressure of the liquid equals the environmental pressure surrounding the liquid. A liquid in a vacuum has a lower boiling point than when that liquid is at atmospheric pressure. A liquid at high-pressure has a higher boiling point than when that liquid is at atmospheric pressure.
Melting point	The melting point of a solid is the temperature at which it changes state from solid to liquid. At the melting point the solid and liquid phase exist in equilibrium. The melting point of a substance depends (usually slightly) on pressure and is usually specified at standard pressure.
Progesterone	Progesterone also known as P4 (pregn-4-ene-3,20-dione) is a C-21 steroid hormone involved in the female menstrual cycle, pregnancy (supports gestation) and embryogenesis of humans and other species. Progesterone belongs to a class of hormones called progestogens, and is the major naturally occurring human progestogen.

Chapter 5. Organic compounds 1: the framework of life

	Progesterone is commonly manufactured from the yam family, Dioscorea.
Steroid	A steroid is a type of organic compound that contains a characteristic arrangement of four cycloalkane rings that are joined to each other. Examples of steroids include the dietary fat cholesterol, the sex hormones estradiol and testosterone, and the anti-inflammatory drug dexamethasone. The core of steroids is composed of twenty carbon atoms bonded together that take the form of four fused rings: three cyclohexane rings (designated as rings A, B, and C in the figure to the right) and one cyclopentane ring (the D ring).
Solubility	Solubility is the property of a solid, liquid, or gaseous chemical substance called solute to dissolve in a solid, liquid, or gaseous solvent to form a homogeneous solution of the solute in the solvent. The solubility of a substance fundamentally depends on the used solvent as well as on temperature and pressure. The extent of the solubility of a substance in a specific solvent is measured as the saturation concentration, where adding more solute does not increase the concentration of the solution.
DNA-binding protein	DNA-binding proteins are proteins that are composed of DNA-binding domains and thus have a specific or general affinity for either single or double stranded DNA. Sequence-specific DNA-binding proteins generally interact with the major groove of B-DNA, because it exposes more functional groups that identify a base pair. However there are some known minor groove DNA-binding ligands such as Netropsin, Distamycin, Hoechst 33258, Pentamidine and others. Examples DNA-binding proteins include transcription factors which modulate the process of transcription, various polymerases, nucleases which cleave DNA molecules, and histones which are involved in chromosome packaging and transcription in the cell nucleus.
Combustion	Combustion or burning is the sequence of exothermic chemical reactions between a fuel and an oxidant accompanied by the production of heat and conversion of chemical species. The release of heat can result in the production of light in the form of either glowing or a flame. Fuels of interest often include organic compounds (especially hydrocarbons) in the gas, liquid or solid phase.
Intramolecular force	An intramolecular force is any force that holds together the atoms making up a molecule or compound. They contain all types of chemical bond. They are stronger than intermolecular forces, which are present between atoms or molecules that are not actually bonded.

Chapter 5. Organic compounds 1: the framework of life

Double bond	A double bond in chemistry is a chemical bond between two chemical elements involving four bonding electrons instead of the usual two. The most common double bond, that between two carbon atoms, can be found in alkenes. Many types of double bonds between two different elements exist, for example in a carbonyl group with a carbon atom and an oxygen atom.
Triple bond	A triple bond in chemistry is a chemical bond between two chemical elements involving six bonding electrons instead of the usual two in a covalent single bond. The most common triple bond, that between two carbon atoms, can be found in alkynes. Other functional groups containing a triple bond are cyanides and isocyanides.
Lewis structure	Lewis structures (also known as Lewis dot diagrams, electron dot diagrams, and electron dot structures) are diagrams that show the bonding between atoms of a molecule and the lone pairs of electrons that may exist in the molecule.
Genetic engineering	Genetic engineering, is the direct human manipulation of an organism's genome using modern DNA technology. It involves the introduction of foreign DNA or synthetic genes into the organism of interest. The introduction of new DNA does not require the use of classical genetic methods, however traditional breeding methods are typically used for the propagation of recombinant organisms.
Geraniol	Geraniol is a monoterpenoid and an alcohol. It is the primary part of rose oil, palmarosa oil, and citronella oil (Java type). It also occurs in small quantities in geranium, lemon, and many other essential oils.
Hormone	A hormone is a chemical released by a cell or a gland in one part of the body that sends out messages that affect cells in other parts of the organism. Only a small amount of hormone is required to alter cell metabolism. In essence, it is a chemical messenger that transports a signal from one cell to another.
Limonene	Limonene is a colourless liquid hydrocarbon classified as a cyclic terpene. The more common D isomer possesses a strong smell of oranges. It is used in chemical synthesis as a precursor to carvone and as a renewably-based solvent in cleaning products.
Epigenetics	In biology, and specifically genetics, epigenetics is the study of changes produced in gene expression caused by mechanisms other than changes in the underlying DNA sequence -hence the name epi- -genetics. Examples of such changes might be DNA methylation or histone acetylation, both of which serve to suppress gene expression without altering the sequence of the silenced genes.

Chapter 5. Organic compounds 1: the framework of life

Gene expression	Gene expression is the process by which information from a gene is used in the synthesis of a functional gene product. These products are often proteins, but in non-protein coding genes such as ribosomal RNA (rRNA), transfer RNA (tRNA) or small nuclear RNA (snRNA) genes, the product is a functional RNA. The process of gene expression is used by all known life - eukaryotes (including multicellular organisms), prokaryotes (bacteria and archaea), possibly induced by viruses - to generate the macromolecular machinery for life. Several steps in the gene expression process may be modulated, including the transcription, RNA splicing, translation, and post-translational modification of a protein.
Histone	In biology, histones are highly alkaline proteins found in eukaryotic cell nuclei that package and order the DNA into structural units called nucleosomes. They are the chief protein components of chromatin, acting as spools around which DNA winds, and play a role in gene regulation. Without histones, the unwound DNA in chromosomes would be very long (a length to width ratio of more than 10 million to one in human DNA).
Methyl group	Methyl group is an alkyl derived from methane, containing one carbon atom bonded to three hydrogen atoms --CH_3. The group is often abbreviated Me. Such hydrocarbon groups occur in many organic compounds.
Acetic acid	Acetic acid ?'si?t?k is an organic compound with the chemical formula CH_3CO_2H (also written as CH_3COOH). It is a colourless liquid that when undiluted is also called glacial acetic acid. Acetic acid is the main component of vinegar (apart from water), and has a distinctive sour taste and pungent smell.
Benzene	Benzene is an organic chemical compound with the molecular formula C_6H_6. Its molecule is composed of 6 carbon atoms joined in a ring, with 1 hydrogen atom attached to each carbon atom. Because its molecules contain only carbon and hydrogen atoms, benzene is classed as a hydrocarbon.
Phenyl group	In organic chemistry, the phenyl group is a cyclic group of atoms with the formula C_6H_5. Phenyl groups are closely related to benzene. Phenyl groups have six carbon atoms bonded together in a hexagonal planar ring, five of which are bonded to individual hydrogen atoms, with the remaining carbon bonded to a substituent.
Electronegativity	Electronegativity, symbol χ, is a chemical property that describes the tendency of an atom or a functional group to attract electrons towards itself. An atom's electronegativity is affected by both its atomic number and the distance that its valence electrons reside from the charged nucleus. The higher the associated electronegativity number, the more an element or compound attracts electrons towards it.

Chapter 5. Organic compounds 1: the framework of life

| Cell membrane | The cell membrane is a biological membrane that separates the interior of all cells from the outside environment. The cell membrane is selectively permeable to ions and organic molecules and controls the movement of substances in and out of cells. It basically protects the cell from outside forces. |

PRACTICE QUIZ
Chapter 5. Organic compounds 1: the framework of life

1. _____ - Environmental problems, chemical approaches - is a peer-reviewed scientific journal publishing original research and review articles on all aspects of environmental chemistry. The Journal publishes manuscripts addressing the chemistry of the environment (air, water, soil, sediments, space, and biota) including atmospheric chemistry, (bio)geochemistry, climate change, marine chemistry, water chemistry, polar chemistry, fire chemistry, astrochemistry, earth and geochemistry, soil and sediment chemistry and chemical toxicology.

   Papers are written in a style that is accessible to those outside the field, as the readership will include chemists, biologists, toxicologists, soil scientists, and workers from government and industrial institutions.

   a. Eutrophication
   b. Exposure assessment
   c. Environmental Chemistry
   d. OTEX Ozone Laundry System

2. The mineral _____ is an allotrope of carbon. It was named by Abraham Gottlob Werner in 1789 from the Ancient Greek γρ?φω (grapho), 'to draw/write', for its use in pencils, where it is commonly called lead (not to be confused with the metallic element lead). Unlike diamond (another carbon allotrope), _____ is an electrical conductor, a semimetal.

   a. Plasma
   b. Graphite
   c. Metallofullerene
   d. Polymer solar cell

3. _____, symbol χ, is a chemical property that describes the tendency of an atom or a functional group to attract electrons towards itself. An atom's _____ is affected by both its atomic number and the distance that its valence electrons reside from the charged nucleus. The higher the associated _____ number, the more an element or compound attracts electrons towards it.

   a. Electronegativity
   b. Inert
   c. Ionization energy
   d. Oxidation state

---

4. _____s (also known as paraffins or saturated hydrocarbons) are chemical compounds that consist only of hydrogen and carbon atoms and are bonded exclusively by single bonds (i.e., they are saturated compounds) without any cycles . _____s belong to a homologous series of organic compounds in which the members differ by a constant relative molecular mass of 14. They have 2 main commercial sources, crude oil and natural gas.

Each carbon atom has 4 bonds (either C-H or C-C bonds), and each hydrogen atom is joined to a carbon atom (H-C bonds).

   a. Alkene
   b. Alkane
   c. Ethylenediamine pyrocatechol
   d. Explosophore

5. _____ 'kɑːrbən is the chemical element with symbol C and atomic number 6. As a member of group 14 on the periodic table, it is nonmetallic and tetravalent–making four electrons available to form covalent chemical bonds. There are three naturally occurring isotopes, with ^{12}C and ^{13}C being stable, while ^{14}C is radioactive, decaying with a half-life of about 5,730 years. _____ is one of the few elements known since antiquity.

   a. CHON
   b. Chromium
   c. Carbon
   d. Fluorine

ANSWER KEY
Chapter 5. Organic compounds 1: the framework of life

1. c
2. b
3. a
4. b
5. c

You can take the complete Chapter Practice Test

for Chapter 5. Organic compounds 1: the framework of life

on all key terms, persons, places, and concepts.

Online 99 Cents

http://www.epub14.51.19910.5.cram101.com/

Use www.Cram101.com for all your study needs

including Cram101's online interactive problem solving labs in chemistry, statistics, mathematics, and more.

CHAPTER OUTLINE: KEY TERMS, PEOPLE, PLACES, CONCEPTS
Chapter 6
Organic compounds 2: adding function to the framework of life

_____ | Alcohol
_____ | Polarity
_____ | Alkoxy group
_____ | Ether
_____ | Solubility
_____ | Metabolism
_____ | Aldehyde
_____ | Ketone
_____ | Electronegativity
_____ | DNA profiling
_____ | Electron pair
_____ | Trivial name
_____ | Carboxylic acid
_____ | Glycolysis
_____ | Avogadro constant
_____ | Anaerobic respiration
_____ | Citric acid
_____ | Citric acid cycle
_____ | Lactic acid

Chapter 6. Organic compounds 2: adding function to the framework of life

- _____ Pyruvic acid
- _____ Fatty acid
- _____ Salicylic acid
- _____ Amine
- _____ Gel electrophoresis
- _____ Boiling point
- _____ Alkaloid
- _____ Caffeine
- _____ Codeine
- _____ Morphine
- _____ Amide
- _____ Amino acid
- _____ Haloalkane
- _____ Halogen
- _____ Natural gas
- _____ Thiol
- _____ Bromoethane
- _____ Chloromethane
- _____ Halothane

Chapter 6. Organic compounds 2: adding function to the framework of life

- Ozone layer
- Fluoromethane

CHAPTER HIGHLIGHTS: KEY TERMS, PEOPLE, PLACES, CONCEPTS
Chapter 6. Organic compounds 2: adding function to the framework of life

Alcohol	In chemistry, an alcohol is an organic compound in which the hydroxyl functional group (-OH) is bound to a carbon atom. In particular, this carbon center should be saturated, having single bonds to three other atoms. An important class of alcohols are the simple acyclic alcohols, the general formula for which is $C_nH_{2n+1}OH$. Of those, ethanol (C_2H_5OH) is the type of alcohol found in alcoholic beverages, and in common speech the word alcohol refers specifically to ethanol.
Polarity	In physics, polarity is a description of an attribute, typically a binary attribute (one with two values), or a vector (a direction). For example: • An electric charge has a polarity of either positive or negative. • A battery contains polarity, with the two + and - terminals. Similar to electric charge, the energy flows from the positive terminal, through the battery, to the negative terminal, and exits the dry cell. • A voltage has a polarity, in that it could be positive or negative (with respect to some other voltage, such as the one at the other end of a battery or electric circuit). • A magnet has a polarity, in that one end is the 'north' and the other is the 'south'. • The spin of an entity in quantum mechanics has a polarity - positive or negative. • Polarized light has waves which all line up in the same direction. Chemical polarity is a feature of chemical bonds, where two different atoms in the same molecule have different electronegativity.
Alkoxy group	In chemistry, the alkoxy group is an alkyl (carbon and hydrogen chain) group singular bonded to oxygen thus: R--O. The range of alkoxy groups is great, the simplest being methoxy (CH_3O--). An ethoxy group (CH_3CH_2O--) is found in the organic compound phenetol, $C_6H_5OCH_2CH_3$ which is also known as ethoxy benzene. Related to alkoxy groups are aryloxy groups, which have an aryl group singular bonded to oxygen such as the phenoxy group (C_6H_5O--).
Ether	Wikimedia.org/wikipedia/commons/thumb/5/51/Ether-%28general%29.png/150px-Ether-%28general%29.png' width='150' height='75' />

Chapter 6. Organic compounds 2: adding function to the framework of life

	Ethers ( ) are a class of organic compounds that contain an ether group -- an oxygen atom connected to two alkyl or aryl groups -- of general formula R-O-R'. A typical example is the solvent and anesthetic diethyl ether, commonly referred to simply as 'ether' (CH_3-CH_2-O-CH_2-CH_3). Ethers are common in organic chemistry and pervasive in biochemistry, as they are common linkages in carbohydrates and lignin.
Solubility	Solubility is the property of a solid, liquid, or gaseous chemical substance called solute to dissolve in a solid, liquid, or gaseous solvent to form a homogeneous solution of the solute in the solvent. The solubility of a substance fundamentally depends on the used solvent as well as on temperature and pressure. The extent of the solubility of a substance in a specific solvent is measured as the saturation concentration, where adding more solute does not increase the concentration of the solution.
Metabolism	Metabolism is the set of chemical reactions that happen in the cells of living organisms to sustain life. These processes allow organisms to grow and reproduce, maintain their structures, and respond to their environments. The word metabolism can also refer to all chemical reactions that occur in living organisms, including digestion and the transport of substances into and between different cells, in which case the set of reactions within the cells is called intermediary metabolism or intermediate metabolism.
Aldehyde	An aldehyde is an organic compound containing a formyl group. This functional group, with the structure R-CHO, consists of a carbonyl center (a carbon double bonded to oxygen) bonded to hydrogen and an R group, which is any generic alkyl or side chain. The group without R is called the aldehyde group or formyl group.
Ketone	In organic chemistry, a ketone is an organic compound with the structure RC(=O)R', where R and R' can be a variety of carbon-containing substituents. It features a carbonyl group (C=O) bonded to two other carbon atoms. The general formula for ketones is $C_nH_{2n}O$. It is the same for Aldehydes; the difference is however in their respective structures- Ketones have two alkyl substituents (Alkyl groups) at each end of their carbonyl group while Aldehydes have one alkyl substituent at one end of their carbonyl group.
Electronegativity	Electronegativity, symbol χ, is a chemical property that describes the tendency of an atom or a functional group to attract electrons towards itself. An atom's electronegativity is affected by both its atomic number and the distance that its valence electrons reside from the charged nucleus. The higher the associated electronegativity number, the more an element or compound attracts electrons towards it.

Chapter 6. Organic compounds 2: adding function to the framework of life

DNA profiling	DNA profiling is a technique employed by forensic scientists to assist in the identification of individuals by their respective DNA profiles. DNA profiles are encrypted sets of numbers that reflect a person's DNA makeup, which can also be used as the person's identifier. DNA profiling should not be confused with full genome sequencing.
Electron pair	In chemistry, an electron pair consists of two electrons that occupy the same orbital but have opposite spins.
	Because electrons are fermions, the Pauli exclusion principle forbids these particles from having exactly the same quantum numbers. Therefore the only way to occupy the same orbital, i.e. have the same orbital quantum numbers, is to differ in the spin quantum number.
Trivial name	In chemistry, a trivial name is a common name or vernacular name; it is a non-systematic name or non-scientific name. That is, the name is not recognized according to the rules of any formal (e.g. IUPAC) system of nomenclature. A limited number of trivial chemical names, however, are retained names and are yet part of the nomenclature.
Carboxylic acid	Carboxylic acids ( ) are organic acids characterized by the presence of at least one carboxyl group. The general formula of a carboxylic acid is R-COOH, where R is some monovalent functional group. A carboxyl group is a functional group consisting of a carbonyl (RR'C=O) and a hydroxyl (R-O-H), which has the formula -C(=O)OH, usually written as -COOH or $-CO_2H$.
	Carboxylic acids are Brønsted-Lowry acids because they are proton (H^+) donors.
Glycolysis	Glycolysis is the metabolic pathway that converts glucose $C_6H_{12}O_6$, into pyruvate, $CH_3COCOO^- + H^+$. The free energy released in this process is used to form the high-energy compounds ATP (adenosine triphosphate), $FADH_2$ and NADH (reduced nicotinamide adenine dinucleotide).
	Glycolysis is a definite sequence of ten reactions involving ten intermediate compounds (one of the steps involves two intermediates).

Chapter 6. Organic compounds 2: adding function to the framework of life

Avogadro constant	In chemistry and physics, the Avogadro constant is defined as the ratio of the number of constituent particles N (usually atoms or molecules) in a sample to the amount of substance n (unit mole) through the relationship $N_A = Nn$. Thus, it is the proportionality factor that relates the molar mass of an entity, i.e., the mass per amount of substance, to the mass of said entity. The Avogadro constant expresses the number of elementary entities per mole of substance and it has the value $6.022\ 141\ 29(27) \times 10^{23}\ mol^{-1}$.
Anaerobic respiration	Anaerobic respiration is a form of respiration using electron acceptors other than oxygen. Although oxygen is not used as the final electron acceptor, the process still uses a respiratory electron transport chain; it is respiration without oxygen. In order for the electron transport chain to function, an exogenous final electron acceptor must be present to allow electrons to pass through the system.
Citric acid	Citric acid is a weak organic acid. It is a natural preservative/conservative and is also used to add an acidic, or sour, taste to foods and soft drinks. In biochemistry, the conjugate base of citric acid, citrate, is important as an intermediate in the citric acid cycle, and therefore occurs in the metabolism of virtually all living things.
Citric acid cycle	The citric acid cycle -- also known as the tricarboxylic acid cycle (TCA cycle), the Krebs cycle, or the Szent-Györgyi-Krebs cycle -- is a series of chemical reactions used by all aerobic organisms to generate energy through the oxidization of acetate derived from carbohydrates, fats and proteins into carbon dioxide and water. In addition, the cycle provides precursors for the biosynthesis of compounds including certain amino acids as well as the reducing agent NADH that is used in numerous biochemical reactions. Its central importance to many biochemical pathways suggests that it was one of the earliest established components of cellular metabolism and may have originated abiogenically.
Lactic acid	Lactic acid, is a chemical compound that plays a role in various biochemical processes and was first isolated in 1780 by the Swedish chemist Carl Wilhelm Scheele. Lactic acid is a carboxylic acid with the chemical formula $C_3H_6O_3$. It has a hydroxyl group adjacent to the carboxyl group, making it an alpha hydroxy acid (AHA).
Pyruvic acid	Pyruvic acid is an organic acid, a ketone, as well as the simplest of the alpha-keto acids. The carboxylate (COO^-) anion of pyruvic acid, its Brønsted-Lowry conjugate base, CH_3COCOO^-, is known as pyruvate, and is a key intersection in several metabolic pathways.

Chapter 6. Organic compounds 2: adding function to the framework of life

	Pyruvate can be made from glucose through glycolysis, converted back to carbohydrates (such as glucose) via gluconeogenesis, or to fatty acids through acetyl-CoA. It can also be used to construct the amino acid alanine and be converted into ethanol.
Fatty acid	In chemistry, especially biochemistry, a fatty acid is a carboxylic acid with a long aliphatic tail (chain), which is either saturated or unsaturated. Most naturally occurring fatty acids have a chain of an even number of carbon atoms, from 4 to 28. Fatty acids are usually derived from triglycerides or phospholipids. When they are not attached to other molecules, they are known as 'free' fatty acids.
Salicylic acid	Salicylic acid is a monohydroxybenzoic acid, a type of phenolic acid and a beta hydroxy acid. This colorless crystalline organic acid is widely used in organic synthesis and functions as a plant hormone. It is derived from the metabolism of salicin.
Amine	Amines are organic compounds and functional groups that contain a basic nitrogen atom with a lone pair. Amines are derivatives of ammonia, wherein one or more hydrogen atoms have been replaced by a substituent such as an alkyl or aryl group. Important amines include amino acids, biogenic amines, trimethylamine, and aniline.
Gel electrophoresis	Gel electrophoresis is a method used in clinical chemistry to separate proteins by charge and or size (IEF agarose, essentially size independent) and in biochemistry and molecular biology to separate a mixed population of DNA and RNA fragments by length, to estimate the size of DNA and RNA fragments or to separate proteins by charge. Nucleic acid molecules are separated by applying an electric field to move the negatively charged molecules through an agarose matrix. Shorter molecules move faster and migrate farther than longer ones because shorter molecules migrate more easily through the pores of the gel.
Boiling point	The boiling point of a substance is the temperature at which the vapor pressure of the liquid equals the environmental pressure surrounding the liquid. A liquid in a vacuum has a lower boiling point than when that liquid is at atmospheric pressure. A liquid at high-pressure has a higher boiling point than when that liquid is at atmospheric pressure.

Chapter 6. Organic compounds 2: adding function to the framework of life

Alkaloid	Alkaloids are a group of naturally occurring chemical compounds that contain mostly basic nitrogen atoms. This group also includes some related compounds with neutral and even weakly acidic properties. Also some synthetic compounds of similar structure are attributed to alkaloids.
Caffeine	Caffeine Caffeine is a bitter, white crystalline xanthine alkaloid that acts as a stimulant drug. Caffeine is found in varying quantities in the seeds, leaves, and fruit of some plants, where it acts as a natural pesticide that paralyzes and kills certain insects feeding on the plants. It is most commonly consumed by humans in infusions extracted from the seed of the coffee plant and the leaves of the tea bush, as well as from various foods and drinks containing products derived from the kola nut.
Codeine	Codeine, the other being the semi-synthetic 6-methylmorphine) is an opiate used for its analgesic, antitussive, and antidiarrheal properties. Codeine is the second-most predominant alkaloid in opium, at up to three percent; it is much more prevalent in the Iranian poppy (Papaver bractreatum), and codeine is extracted from this species in some places although the below-mentioned morphine methylation process is still much more common. It is considered the prototype of the weak to midrange opioids (tramadol, dextropropoxyphene, dihydrocodeine, hydrocodone).
Morphine	Morphine (; MS Contin, MSIR, Avinza, Kadian, Oramorph, Roxanol, Kapanol) is a potent opiate analgesic drug that is used to relieve severe pain. It was first isolated in 1804 by Friedrich Sertürner, first distributed by him in 1817, and first commercially sold by Merck in 1827, which at the time was a single small chemists' shop. It was more widely used after the invention of the hypodermic needle in 1857. It took its name from the Greek god of dreams Morpheus .
Amide	Amide refers to compounds with the functional group $R_nE(O)_xNR'_2$ (R and R' refer to H or organic groups). Most common are 'organic amides' (n = 1, E = C, x = 1), but many other important types of amides are known including phosphor amides (n = 2, E = P, x = 1 and many related formulas) and sulfonamides (E = S, x= 2). The term amide refers both to classes of compounds and to the functional group $(R_nE(O)_xNR'_2)$ within those compounds.
Amino acid	Amino acids (?'mi?no?, ?'ma?o?, or 'æm?o?) are molecules containing an amine group, a carboxylic acid group, and a side-chain that is specific to each amino acid. The key elements of an amino acid are carbon, hydrogen, oxygen, and nitrogen. They are particularly important in biochemistry, where the term usually refers to alpha-amino acids.

Chapter 6. Organic compounds 2: adding function to the framework of life

Haloalkane	The haloalkanes (also known as halogenoalkanes or alkyl halides) are a group of chemical compounds derived from alkanes containing one or more halogens. They are a subset of the general class of halocarbons, although the distinction is not often made. Haloalkanes are widely used commercially and, consequently, are known under many chemical and commercial names.
Halogen	The halogens or halogen elements are a series of nonmetal elements from Group 17 IUPAC Style (formerly: VII, VIIA) of the periodic table, comprising fluorine (F), chlorine (Cl), bromine (Br), iodine (I), and astatine (At). The artificially created element 117, provisionally referred to by the systematic name ununseptium, may also be a halogen. The group of halogens is the only periodic table group which contains elements in all three familiar states of matter at standard temperature and pressure.
Natural gas	Natural gas is a naturally occurring hydrocarbon gas mixture consisting primarily of methane, with up to 20 % of other hydrocarbons as well as impurities in varying amounts such as carbon dioxide. Natural gas is widely used as an important energy source in many applications including heating buildings, generating electricity, providing heat and power to industry, as fuel for vehicles and as a chemical feedstock in the manufacture of products such as plastics and other commercially important organic chemicals. Natural gas is found in deep underground natural rock formations or associated with other hydrocarbon reservoirs, in coal beds, and as methane clathrates.
Thiol	In organic chemistry, a thiol is an organosulfur compound that contains a carbon-bonded sulfhydryl (-C-SH or R-SH) group (where R represents an alkane, alkene, or other carbon-containing group of atoms). Thiols are the sulfur analogue of alcohols (that is, sulfur takes the place of oxygen in the hydroxyl group of an alcohol), and the word is a portmanteau of 'thio' + 'alcohol,' with the first word deriving from Greek θε?ον ('thion') = 'sulfur'. The -SH functional group itself is referred to as either a thiol group or a sulfhydryl group.
Bromoethane	Bromoethane, is a chemical compound of the haloalkanes group. It is abbreviated by chemists as EtBr. This volatile compound has an ether-like odour.

Chapter 6. Organic compounds 2: adding function to the framework of life

Chloromethane	Chloromethane, R-40 or HCC 40, is a chemical compound of the group of organic compounds called haloalkanes. It was once widely used as a refrigerant. It is a colorless extremely flammable gas with a mildly sweet odor, which is, however, detected at possibly toxic levels.
Halothane	Halothane is an inhalational general anesthetic. Its IUPAC name is 2-bromo-2-chloro-1,1,1-trifluoroethane. It is the only inhalational anesthetic agent containing a bromine atom; there are several other halogenated anesthesia agents which lack the bromine atom and do contain the fluorine and chlorine atoms present in halothane.
Ozone layer	The ozone layer is a layer in Earth's atmosphere containing relatively high concentrations of ozone (O_3). However, 'relatively high,' in the case of ozone, is still very small with regard to ordinary oxygen, and is less than 10 parts per million, with the average ozone concentration in Earth's atmosphere being only about 0.6 parts per million. The ozone layer is mainly located in the lower portion of the stratosphere from approximately 20 to 30 kilometres (12 to 19 mi) above Earth, though the thickness varies seasonally and geographically.
Fluoromethane	Fluoromethane, Freon 41, Halocarbon-41 and HFC-41, is a non-toxic, liquefiable, and flammable gas at standard temperature and pressure. It is made of carbon, hydrogen, and fluorine. The name stems from the fact that it is methane (CH_4) plus fluorine, minus a hydrogen.

PRACTICE QUIZ
Chapter 6. Organic compounds 2: adding function to the framework of life

1. _____s (?'mi?no?, ?'ma?o?, or 'æm?o?) are molecules containing an amine group, a carboxylic acid group, and a side-chain that is specific to each _____. The key elements of an _____ are carbon, hydrogen, oxygen, and nitrogen. They are particularly important in biochemistry, where the term usually refers to alpha-_____s.

    a. Arsenobetaine
    b. Amino acid
    c. Isoionic point
    d. Azide

2. _____ refers to compounds with the functional group $R_nE(O)_xNR'_2$ (R and R' refer to H or organic groups). Most common are 'organic _____s' (n = 1, E = C, x = 1), but many other important types of _____s are known including phosphor_____s (n = 2, E = P, x = 1 and many related formulas) and sulfonamides (E = S, x= 2). The term _____ refers both to classes of compounds and to the functional group ($R_nE(O)_xNR'_2$) within those compounds.

    a. Amidine
    b. Amide
    c. Amine oxide
    d. Azide

3. An _____ is an organic compound containing a formyl group. This functional group, with the structure R-CHO, consists of a carbonyl center (a carbon double bonded to oxygen) bonded to hydrogen and an R group, which is any generic alkyl or side chain. The group without R is called the _____ group or formyl group.

    a. Acetaldehyde
    b. Acrolein
    c. Aldol
    d. Aldehyde

4. Wikimedia.org/wikipedia/commons/thumb/5/51/_____-%28general%29.png/150px-_____-%28general%29.png' width='150' height='75' />

    _____s ( ) are a class of organic compounds that contain an _____ group -- an oxygen atom connected to two alkyl or aryl groups -- of general formula R-O-R'. A typical example is the solvent and anesthetic diethyl _____, commonly referred to simply as '_____' (CH_3-CH_2-O-CH_2-CH_3). _____s are common in organic chemistry and pervasive in biochemistry, as they are common linkages in carbohydrates and lignin.

---

a. Imine
   b. Ether
   c. Isothiouronium
   d. Orthoester

5. _____ is a naturally occurring hydrocarbon gas mixture consisting primarily of methane, with up to 20 % of other hydrocarbons as well as impurities in varying amounts such as carbon dioxide. _____ is widely used as an important energy source in many applications including heating buildings, generating electricity, providing heat and power to industry, as fuel for vehicles and as a chemical feedstock in the manufacture of products such as plastics and other commercially important organic chemicals.

   _____ is found in deep underground natural rock formations or associated with other hydrocarbon reservoirs, in coal beds, and as methane clathrates.

   a. Renewable natural gas
   b. Billion cubic metres of natural gas
   c. Gas to liquids
   d. Natural gas

ANSWER KEY
Chapter 6. Organic compounds 2: adding function to the framework of life

1. b
2. b
3. d
4. b
5. d

You can take the complete Chapter Practice Test

for Chapter 6. Organic compounds 2: adding function to the framework of life

on all key terms, persons, places, and concepts.

Online 99 Cents

http://www.epub14.51.19910.6.cram101.com/

Use www.Cram101.com for all your study needs

including Cram101's online interactive problem solving labs in chemistry, statistics, mathematics, and more.

CHAPTER OUTLINE: KEY TERMS, PEOPLE, PLACES, CONCEPTS
Chapter 7
Biological macromolecules: providing life's infrastructure

- _____ Amino acid
- _____ Ionization
- _____ Polymerization
- _____ Polymer
- _____ Avogadro constant
- _____ Glutamic acid
- _____ Insulin
- _____ Monosodium glutamate
- _____ Peptide
- _____ Proinsulin
- _____ Aspartame
- _____ Phenylalanine
- _____ Phenylketonuria
- _____ Carbohydrate
- _____ Monosaccharide
- _____ Photosynthesis
- _____ Polysaccharide
- _____ Racemic mixture
- _____ Respiration

Chapter 7. Biological macromolecules: providing life`s infrastructure

_____ DNA profiling

_____ Aldose

_____ Ketose

_____ Lipid

_____ Steroid

_____ Endocrine system

_____ Cholesterol

_____ Fatty acid

_____ High-density lipoprotein

_____ Lipoprotein

_____ Lone pair

_____ Low-density lipoprotein

_____ Cell membrane

_____ Glycerophospholipid

_____ Lipid bilayer

_____ Trans fat

_____ Unsaturated fat

_____ Amino alcohol

_____ Choline

Chapter 7. Biological macromolecules: providing life's infrastructure

- Ethanolamine
- Gel electrophoresis
- Serine
- Nucleic acid
- Transmembrane protein
- Cytosine
- Guanine
- Nitrogenous base
- Pyrimidine
- Hydrophobicity
- Human genome
- Human Genome Project
- Adenosine diphosphate
- Adenosine monophosphate
- Sequencing
- Covalent bond
- Zinc finger
- DNA-binding protein
- Carbonic anhydrase

Chapter 7. Biological macromolecules: providing life`s infrastructure

- Metalloprotein
- Gasotransmitter
- Nitric oxide

CHAPTER HIGHLIGHTS: KEY TERMS, PEOPLE, PLACES, CONCEPTS
Chapter 7. Biological macromolecules: providing life's infrastructure

Amino acid	Amino acids (?'mi?no?, ?'ma?o?, or 'æm?o?) are molecules containing an amine group, a carboxylic acid group, and a side-chain that is specific to each amino acid. The key elements of an amino acid are carbon, hydrogen, oxygen, and nitrogen. They are particularly important in biochemistry, where the term usually refers to alpha-amino acids.
Ionization	Ionization is the process of converting an atom or molecule into an ion by adding or removing charged particles such as electrons or ions. In the case of ionisation of a gas, ion-pairs are created consisting of a free electron and a +ve ion. Types of Ionisation The process of ionization works slightly differently depending on whether an ion with a positive or a negative electric charge is being produced.
Polymerization	In polymer chemistry, polymerization is a process of reacting monomer molecules together in a chemical reaction to form polymer chains or three-dimensional networks. There are many forms of polymerization and different systems exist to categorize them. In chemical compounds, polymerization occurs via a variety of reaction mechanisms that vary in complexity due to functional groups present in reacting compounds and their inherent steric effects explained by VSEPR Theory.
Polymer	A polymer is a large molecule (macromolecule) composed of repeating structural units. These sub-units are typically connected by covalent chemical bonds. Although the term polymer is sometimes taken to refer to plastics, it actually encompasses a large class of compounds comprising both natural and synthetic materials with a wide variety of properties.
Avogadro constant	In chemistry and physics, the Avogadro constant is defined as the ratio of the number of constituent particles N (usually atoms or molecules) in a sample to the amount of substance n (unit mole) through the relationship $N_A = Nn$. Thus, it is the proportionality factor that relates the molar mass of an entity, i.e., the mass per amount of substance, to the mass of said entity. The Avogadro constant expresses the number of elementary entities per mole of substance and it has the value $6.022\ 141\ 29(27) \times 10^{23}\ \text{mol}^{-1}$.

Chapter 7. Biological macromolecules: providing life's infrastructure

Glutamic acid	Glutamic acid is one of the 20-22 proteinogenic amino acids, and its codons are GAA and GAG. It is a non-essential amino acid. The carboxylate anions and salts of glutamic acid are known as glutamates. In neuroscience, glutamate is an important neurotransmitter that plays a key role in long-term potentiation and is important for learning and memory.
Insulin	Insulin is a hormone, produced by the pancreas, which is central to regulating carbohydrate and fat metabolism in the body. Insulin causes cells in the liver, muscle, and fat tissue to take up glucose from the blood, storing it as glycogen inside these tissues. Insulin stops the use of fat as an energy source by inhibiting the release of glucagon.
Monosodium glutamate	Monosodium glutamate, is the sodium salt of glutamic acid, one of the most abundant naturally occurring non-essential amino acids. It was classified by the U.S. Food and Drug Administration as generally recognized as safe (GRAS) and by the European Union as a food additive. MSG has the HS code 29224220 and the E number E621. The glutamate of MSG confers the same umami taste of glutamate from other foods, being chemically identical.
Peptide	Peptides are short polymers of amino acid monomers linked by peptide bonds. They are distinguished from proteins on the basis of size, typically containing fewer than 50 monomer units. The shortest peptides are dipeptides, consisting of two amino acids joined by a single peptide bond.
Proinsulin	Proinsulin is the prohormone precursor to insulin made in the beta cells of the islets of Langerhans, specialized regions of the pancreas. In humans, proinsulin is encoded by the INS gene. Synthesis and post-translational modification Proinsulin is synthesized in the endoplasmic reticulum, where it is folded and its disulfide bonds are oxidized.
Aspartame	Aspartame is an artificial, non-saccharide sweetener used as a sugar substitute in some foods and beverages. In the European Union, it is codified as E951. Aspartame is a methyl ester of the aspartic acidphenylalanine dipeptide. It was first sold under the brand name NutraSweet; since 2009 it also has been sold under the brand name AminoSweet.

Chapter 7. Biological macromolecules: providing life's infrastructure

Phenylalanine	Phenylalanine is an α-amino acid with the formula $HO_2CCH(NH_2)CH_2C_6H_5$. This essential amino acid is classified as nonpolar because of the hydrophobic nature of the benzyl side chain. L-Phenylalanine is an electrically neutral amino acid, one of the twenty common amino acids used to biochemically form proteins, coded for by DNA. The codons for L-phenylalanine are UUU and UUC. Phenylalanine is a precursor for tyrosine, the monoamine signaling molecules dopamine, norepinephrine (noradrenaline), and epinephrine (adrenaline), and the skin pigment melanin.
Phenylketonuria	Phenylketonuria is an autosomal recessive metabolic genetic disorder characterized by an error in the genetic code for the hepatic enzyme phenylalanine hydroxylase (PAH), rendering it nonfunctional. This enzyme is necessary to metabolize the amino acid phenylalanine (Phe) to the amino acid tyrosine. When PAH enzymatic activity is reduced, phenylalanine accumulates and is converted into phenylpyruvate (also known as phenylketone), which is detected in the urine.
Carbohydrate	A carbohydrate is an organic compound that consists only of carbon, hydrogen, and oxygen, usually with a hydrogen:oxygen atom ratio of 2:1 (as in water); in other words, with the empirical formula $C_m(H_2O)_n$. (Some exceptions exist; for example, deoxyribose, a component of DNA, has the empirical formula $C_5H_{10}O_4$). Carbohydrates are not technically hydrates of carbon.
Monosaccharide	Monosaccharides are the most basic units of biologically important carbohydrates. They are the simplest form of sugar and are usually colorless, water-soluble, crystalline solids. Some monosaccharides have a sweet taste.
Photosynthesis	Photosynthesis is a process used by plants and other organisms to capture the sun's energy to split off water's hydrogen from oxygen. Hydrogen is combined with carbon dioxide (absorbed from air or water) to form glucose and release oxygen. All living cells in turn use fuels derived from glucose and oxidize the hydrogen and carbon to release the sun's energy and reform water and carbon dioxide in the process (cellular respiration).
Polysaccharide	Polysaccharides are long carbohydrate molecules of repeated monomer units joined together by glycosidic bonds. They range in structure from linear to highly branched. Polysaccharides are often quite heterogeneous, containing slight modifications of the repeating unit.
Racemic mixture	In chemistry, a racemic mixture, is one that has equal amounts of left- and right-handed enantiomers of a chiral molecule. The first known racemic mixture was 'racemic acid', which Louis Pasteur found to be a mixture of the two enantiomeric isomers of tartaric acid. Nomenclature

Chapter 7. Biological macromolecules: providing life's infrastructure

	A racemic mixture is denoted by the prefix (±)- or dl- (for sugars the prefix DL- may be used), indicating an equal (1:1) mixture of dextro and levo isomers.
Respiration	In physiology, respiration (often confused with breathing) is defined as the transport of oxygen from the outside air to the cells within tissues, and the transport of carbon dioxide in the opposite direction. This is in contrast to the biochemical definition of respiration, which refers to cellular respiration: the metabolic process by which an organism obtains energy by reacting oxygen with glucose to give water, carbon dioxide and ATP (energy). Although physiologic respiration is necessary to sustain cellular respiration and thus life in animals, the processes are distinct: cellular respiration takes place in individual cells of the organism, while physiologic respiration concerns the bulk flow and transport of metabolites between the organism and the external environment.
DNA profiling	DNA profiling is a technique employed by forensic scientists to assist in the identification of individuals by their respective DNA profiles. DNA profiles are encrypted sets of numbers that reflect a person's DNA makeup, which can also be used as the person's identifier. DNA profiling should not be confused with full genome sequencing.
Aldose	An aldose is a monosaccharide (a simple sugar) that contains only one aldehyde (-CH=O) group per molecule. The chemical formula takes the form $C_n(H_2O)_n$. The simplest possible aldose is the diose glycolaldehyde, which only contains two carbon atoms.
Ketose	A ketose is a sugar containing one ketone group per molecule. With three carbon atoms, dihydroxyacetone is the simplest of all ketoses and is the only one having no optical activity. Ketoses can isomerize into an aldose when the carbonyl group is located at the end of the molecule.
Lipid	Lipids constitute a broad group of naturally occurring molecules that include fats, waxes, sterols, fat-soluble vitamins (such as vitamins A, D, E, and K), monoglycerides, diglycerides, triglycerides, phospholipids, and others. The main biological functions of lipids include energy storage, as structural components of cell membranes, and as important signaling molecules.

Chapter 7. Biological macromolecules: providing life's infrastructure

Lipids may be broadly defined as hydrophobic or amphiphilic small molecules; the amphiphilic nature of some lipids allows them to form structures such as vesicles, liposomes, or membranes in an aqueous environment.

Steroid	A steroid is a type of organic compound that contains a characteristic arrangement of four cycloalkane rings that are joined to each other. Examples of steroids include the dietary fat cholesterol, the sex hormones estradiol and testosterone, and the anti-inflammatory drug dexamethasone. The core of steroids is composed of twenty carbon atoms bonded together that take the form of four fused rings: three cyclohexane rings (designated as rings A, B, and C in the figure to the right) and one cyclopentane ring (the D ring).
Endocrine system	The endocrine system is the system of glands, each of which secretes a type of hormone directly into the bloodstream to regulate the body. The endocrine system is in contrast to the exocrine system, which secretes its chemicals using ducts. It derives from the Greek words 'endo' meaning inside, within, and 'crinis' for secrete.
Cholesterol	Cholesterol is an organic chemical substance classified as a waxy steroid of fat. It is an essential structural component of mammalian cell membranes and is required to establish proper membrane permeability and fluidity. In addition to its importance within cells, cholesterol is an important component in the hormonal systems of the body for the manufacture of bile acids, steroid hormones, and vitamin D. Cholesterol is the principal sterol synthesized by animals; in vertebrates it is formed predominantly in the liver.
Fatty acid	In chemistry, especially biochemistry, a fatty acid is a carboxylic acid with a long aliphatic tail (chain), which is either saturated or unsaturated. Most naturally occurring fatty acids have a chain of an even number of carbon atoms, from 4 to 28. Fatty acids are usually derived from triglycerides or phospholipids. When they are not attached to other molecules, they are known as 'free' fatty acids.

Chapter 7. Biological macromolecules: providing life's infrastructure

High-density lipoprotein	High-density lipoprotein is one of the five major groups of lipoproteins, which, in order of sizes, largest to smallest, are chylomicrons, VLDL, IDL, LDL, and HDL, which enable lipids like cholesterol and triglycerides to be transported within the water-based bloodstream. In healthy individuals, about thirty percent of blood cholesterol is carried by HDL.
	Blood tests typically report HDL-C level, i.e. the amount of cholesterol contained in HDL particles. It is often contrasted with low-density or LDL cholesterol or LDL-C. HDL particles are able to remove cholesterol from within artery atheroma and transport it back to the liver for excretion or re-utilization, which is the main reason why the cholesterol carried within HDL particles (HDL-C) is sometimes called 'good cholesterol' (despite the fact that it is exactly the same as the cholesterol in LDL particles).
Lipoprotein	A lipoprotein is a biochemical assembly that contains both proteins and lipids, bound to the proteins, which allow fats to move through the water inside and outside cells. The proteins serve to emulsify the lipid (otherwise called fat) molecules. Many enzymes, transporters, structural proteins, antigens, adhesins, and toxins are lipoproteins.
Lone pair	In chemistry, a lone pair is a valence electron pair without bonding or sharing with other atoms. They are found in the outermost electron shell of an atom, so lone pairs are a subset of a molecule's valence electrons. They can be identified by examining the outermost energy level of an atom--lone electron pairs consist of paired electrons as opposed to single electrons, which may appear if the atomic orbital is not full.
Low-density lipoprotein	Low-density lipoprotein is one of the five major groups of lipoproteins, which in order of size, largest to smallest, are chylomicrons, VLDL, IDL, LDL, and HDL, that enable transport of multiple different fat molecules, including cholesterol, within the water around cells and within the water-based bloodstream. Studies have shown that higher levels of type-B LDL particles (as opposed to type-A LDL particles) promote health problems and cardiovascular disease, they are often informally called the bad cholesterol particles, (as opposed to HDL particles, which are frequently referred to as good cholesterol or healthy cholesterol particles).
	Testing Blood tests typically report LDL-C, the amount of cholesterol contained in LDL. In clinical context, mathematically calculated estimates of LDL-C are commonly used to estimate how much low density lipoproteins are driving progression of atherosclerosis.

Chapter 7. Biological macromolecules: providing life's infrastructure

Cell membrane	The cell membrane is a biological membrane that separates the interior of all cells from the outside environment. The cell membrane is selectively permeable to ions and organic molecules and controls the movement of substances in and out of cells. It basically protects the cell from outside forces.
Glycerophospholipid	Glycerophospholipids or phosphoglycerides are glycerol-based phospholipids. They are the main component of biological membranes. Structures The term glycerophospholipid signifies any derivative of sn-glycero-3-phosphoric acid that contains at least one O-acyl, or O-alkyl, or O-alk-1'-enyl residue attached to the glycerol moiety and a polar head made of a nitrogenous base, a glycerol or an inositol unit.
Lipid bilayer	The lipid bilayer is a thin polar membrane made of two layers of lipid molecules. These membranes are flat sheets that form a continuous barrier around cells. The cell membrane of almost all living organisms and many viruses are made of a lipid bilayer, as are the membranes surrounding the cell nucleus and other sub-cellular structures.
Trans fat	Trans fat is the common name for unsaturated fat with trans-isomer (E-isomer) fatty acid(s). Because the term refers to the configuration of a double carbon-carbon bond, trans fats are sometimes monounsaturated or polyunsaturated, but never saturated. Trans fats do exist in nature but occur far more often during the processing of polyunsaturated fatty acids in food production.
Unsaturated fat	An unsaturated fat is a fat or fatty acid in which there is at least one double bond within the fatty acid chain. A fat molecule is monounsaturated if it contains one double bond, and polyunsaturated if it contains more than one double bond. Where double bonds are formed, hydrogen atoms are eliminated.

Chapter 7. Biological macromolecules: providing life`s infrastructure

Amino alcohol	Amino alcohols are organic compounds that contain both an amine functional group and an alcohol functional group. Common amino alcohols - Ethanolamines - Heptaminol - Isoetarine - Norepinephrine - Propanolamines - Sphingosine - Methanolamine (simplest amino alcohol)
Choline	Choline is a water-soluble essential nutrient. It is usually grouped within the B-complex vitamins. Choline generally refers to the various quaternary ammonium salts containing the N,N,N-trimethylethanolammonium cation.
Ethanolamine	Ethanolamine, is an organic chemical compound that is both a primary amine and a primary alcohol (due to a hydroxyl group). Like other amines, monoethanolamine acts as a weak base. Ethanolamine is a toxic, flammable, corrosive, colorless, viscous liquid with an odor similar to that of ammonia.
Gel electrophoresis	Gel electrophoresis is a method used in clinical chemistry to separate proteins by charge and or size (IEF agarose, essentially size independent) and in biochemistry and molecular biology to separate a mixed population of DNA and RNA fragments by length, to estimate the size of DNA and RNA fragments or to separate proteins by charge. Nucleic acid molecules are separated by applying an electric field to move the negatively charged molecules through an agarose matrix. Shorter molecules move faster and migrate farther than longer ones because shorter molecules migrate more easily through the pores of the gel.
Serine	Serine is an amino acid with the formula $HO_2CCH(NH_2)CH_2OH$. It is one of the proteinogenic amino acids. By virtue of the hydroxyl group, serine is classified as a polar amino acid. Occurrence and biosynthesis

---

Chapter 7. Biological macromolecules: providing life's infrastructure

	This compound is one of the naturally occurring proteinogenic amino acids.
Nucleic acid	Nucleic acids are biological molecules essential for known forms of life on this planet; they include DNA (deoxyribonucleic acid) and RNA (ribonucleic acid). Together with proteins, nucleic acids are the most important biological macromolecules; each is found in abundance in all living things, where they function in encoding, transmitting and expressing genetic information. Nucleic acids were discovered by Friedrich Miescher in 1869. Experimental studies of nucleic acids constitute a major part of modern biological and medical research, and form a foundation for genome and forensic science, as well as the biotechnology and pharmaceutical industries.
Transmembrane protein	A transmembrane protein is a protein that goes from one side of a membrane through to the other side of the membrane. Many TPs function as gateways or 'loading docks' to deny or permit the transport of specific substances across the biological membrane, to get into the cell, or out of the cell as in the case of waste byproducts. As a response to the shape of certain molecules these 'freight handling' TPs may have special ways of folding up or bending that will move a substance through the biological membrane.
Cytosine	Cytosine is one of the four main bases found in DNA and RNA, along with adenine, guanine, and thymine (uracil in RNA). It is a pyrimidine derivative, with a heterocyclic aromatic ring and two substituents attached (an amine group at position 4 and a keto group at position 2). The nucleoside of cytosine is cytidine.
Guanine	Guanine is one of the four main nucleobases found in the nucleic acids DNA and RNA, the others being adenine, cytosine, and thymine (uracil in RNA). In DNA, guanine is paired with cytosine. With the formula $C_5H_5N_5O$, guanine is a derivative of purine, consisting of a fused pyrimidine-imidazole ring system with conjugated double bonds.
Nitrogenous base	A nitrogenous (nitrogen-containing) base is a nitrogen-containing molecule having the chemical properties of a base. It is an organic compound that owes its property as a base to the lone pair of electrons of a nitrogen atom. In biological sciences, nitrogenous bases are typically classified as the derivatives of two parent compounds, pyrimidine and purine.

Chapter 7. Biological macromolecules: providing life's infrastructure

Pyrimidine	Pyrimidine is a heterocyclic aromatic organic compound similar to benzene and pyridine, containing two nitrogen atoms at positions 1 and 3 of the six-member ring. It is isomeric with two other forms of diazine: Pyridazine, with the nitrogen atoms in positions 1 and 2; and Pyrazine, with the nitrogen atoms in positions 1 and 4.

Chemical properties
A pyrimidine has many properties in common with pyridine, as the number of nitrogen atoms in the ring increases the ring pi electrons become less energetic and electrophilic aromatic substitution gets more difficult while nucleophilic aromatic substitution gets easier. |
| Hydrophobicity | In chemistry, hydrophobicity is the physical property of a molecule (known as a hydrophobe) that is repelled from a mass of water.

Hydrophobic molecules tend to be non-polar and thus prefer other neutral molecules and non-polar solvents. Hydrophobic molecules in water often cluster together forming micelles. |
| Human genome | The human (Homo sapiens) genome is stored on 23 chromosome pairs and in the small mitochondrial DNA. Twenty-two of the 23 chromosomes belong to autosomal chromosome pairs, while the remaining pair is sex determinative. The haploid human genome occupies a total of just over three billion DNA base pairs. The Human Genome Project (HGP) produced a reference sequence of the euchromatic human genome and which is used worldwide in the biomedical sciences. |
| Human Genome Project | The Human Genome Project is an international scientific research project with a primary goal of determining the sequence of chemical base pairs which make up DNA, and of identifying and mapping the approximately 20,000-25,000 genes of the human genome from both a physical and functional standpoint.

The project began in October 1990 and was initially headed by Ari Patrinos, head of the Office of Biological and Environmental Research in the U.S. Department of Energy's Office of Science. Francis Collins directed the National Institutes of Health National Human Genome Research Institute efforts. |

Chapter 7. Biological macromolecules: providing life's infrastructure

Adenosine diphosphate	Adenosine diphosphate, abbreviated ADP, is a nucleoside diphosphate. It is an ester of pyrophosphoric acid with the nucleoside adenosine. ADP consists of the pyrophosphate group, the pentose sugar ribose, and the nucleobase adenine.
Adenosine monophosphate	Adenosine monophosphate also known as 5'-adenylic acid, is a nucleotide that is found in RNA. It is an ester of phosphoric acid and the nucleoside adenosine. AMP consists of a phosphate group, the sugar ribose, and the nucleobase adenine. As a substituent it takes the form of the prefix adenylyl-.
Sequencing	In genetics and biochemistry, sequencing means to determine the primary structure (sometimes falsely called primary sequence) of an unbranched biopolymer. Sequencing results in a symbolic linear depiction known as a sequence which succinctly summarizes much of the atomic-level structure of the sequenced molecule. DNA sequencing DNA sequencing is the process of determining the nucleotide order of a given DNA fragment.
Covalent bond	A covalent bond is a form of chemical bonding that is characterized by the sharing of pairs of electrons between atoms. The stable balance of attractive and repulsive forces between atoms when they share electrons is known as covalent bonding. Covalent bonding includes many kinds of interaction, including σ-bonding, π-bonding, metal-to-metal bonding, agostic interactions, and three-center two-electron bonds.
Zinc finger	Zinc fingers are small protein structural motifs that can coordinate one or more zinc ions to help stabilize their folds. They can be classified into several different structural families (zinc finger proteins) and typically function as interaction modules that bind DNA, RNA, proteins, or small molecules. The name 'zinc finger' was originally coined to describe the finger-like appearance of a diagram showing the hypothesized structure of the repeated unit in Xenopus laevis transcription factor IIIA. Classes Zinc fingers coordinate zinc ions with a combination of cysteine and histidine residues.

Chapter 7. Biological macromolecules: providing life's infrastructure

DNA-binding protein	DNA-binding proteins are proteins that are composed of DNA-binding domains and thus have a specific or general affinity for either single or double stranded DNA. Sequence-specific DNA-binding proteins generally interact with the major groove of B-DNA, because it exposes more functional groups that identify a base pair. However there are some known minor groove DNA-binding ligands such as Netropsin, Distamycin, Hoechst 33258, Pentamidine and others. Examples DNA-binding proteins include transcription factors which modulate the process of transcription, various polymerases, nucleases which cleave DNA molecules, and histones which are involved in chromosome packaging and transcription in the cell nucleus.
Carbonic anhydrase	Eukaryotic-type carbonic anhydrase The carbonic anhydrases form a family of enzymes that catalyze the rapid interconversion of carbon dioxide and water to bicarbonate and protons , a reversible reaction that occurs rather slowly in the absence of a catalyst. The active site of most carbonic anhydrases contains a zinc ion; they are therefore classified as metalloenzymes. One of the functions of the enzyme in animals is to interconvert carbon dioxide and bicarbonate to maintain acid-base balance in blood and other tissues, and to help transport carbon dioxide out of tissues.
Metalloprotein	Metalloprotein is a generic term for a protein that contains a metal ion cofactor. A large fraction of all proteins are members of this category, so the area is very large. Metalloenzymes are widespread and diverse in function It is estimated that approximately half of all proteins contain a metal.
Gasotransmitter	Gasotransmitters are gaseous molecules synthesized in the body. They include nitric oxide, hydrogen sulfide, carbon monoxide, and possibly nitrous oxide. Overview

---

Chapter 7. Biological macromolecules: providing life's infrastructure

Gasotransmitters is a family of endogenous molecules of gases or gaseous signaling molecules, including NO, CO, H_2S, and others.

Nitric oxide

Nitric oxide, is a molecule with chemical formula NO. It is a free radical and is an important intermediate in the chemical industry. Nitric oxide is a by-product of combustion of substances in the air, as in automobile engines, fossil fuel power plants, and is produced naturally during the electrical discharges of lightning in thunderstorms.

In mammals including humans, NO is an important cellular signaling molecule involved in many physiological and pathological processes.

PRACTICE QUIZ
Chapter 7. Biological macromolecules: providing life's infrastructure

1. _____ is one of the five major groups of lipoproteins, which, in order of sizes, largest to smallest, are chylomicrons, VLDL, IDL, LDL, and HDL, which enable lipids like cholesterol and triglycerides to be transported within the water-based bloodstream. In healthy individuals, about thirty percent of blood cholesterol is carried by HDL.

   Blood tests typically report HDL-C level, i.e. the amount of cholesterol contained in HDL particles. It is often contrasted with low-density or LDL cholesterol or LDL-C. HDL particles are able to remove cholesterol from within artery atheroma and transport it back to the liver for excretion or re-utilization, which is the main reason why the cholesterol carried within HDL particles (HDL-C) is sometimes called 'good cholesterol' (despite the fact that it is exactly the same as the cholesterol in LDL particles).

   a. Lipid-anchored protein
   b. Lipoprotein-X
   c. High-density lipoprotein
   d. Butyric acid

2. _____ is one of the four main nucleobases found in the nucleic acids DNA and RNA, the others being adenine, cytosine, and thymine (uracil in RNA). In DNA, _____ is paired with cytosine. With the formula $C_5H_5N_5O$, _____ is a derivative of purine, consisting of a fused pyrimidine-imidazole ring system with conjugated double bonds.

   a. Hoelite
   b. Kratochvilite
   c. Guanine
   d. Leonardite

3. A nitrogenous (nitrogen-containing) base is a nitrogen-containing molecule having the chemical properties of a base. It is an organic compound that owes its property as a base to the lone pair of electrons of a nitrogen atom. In biological sciences, _____s are typically classified as the derivatives of two parent compounds, pyrimidine and purine.

   a. S-Nitrosoglutathione
   b. Nitrogenous base
   c. Nucleotide Pyrophosphatase/Phosphodiesterase
   d. Percoll

---

4. A _____ is a biochemical assembly that contains both proteins and lipids, bound to the proteins, which allow fats to move through the water inside and outside cells. The proteins serve to emulsify the lipid (otherwise called fat) molecules. Many enzymes, transporters, structural proteins, antigens, adhesins, and toxins are _____s.

   a. Lipoprotein
   b. Lipotoxicity
   c. Lysochrome
   d. Lysophosphatidylcholine

5. _____ is an α-amino acid with the formula $HO_2CCH(NH_2)CH_2C_6H_5$. This essential amino acid is classified as nonpolar because of the hydrophobic nature of the benzyl side chain. L-_____ is an electrically neutral amino acid, one of the twenty common amino acids used to biochemically form proteins, coded for by DNA. The codons for L-_____ are UUU and UUC. _____ is a precursor for tyrosine, the monoamine signaling molecules dopamine, norepinephrine (noradrenaline), and epinephrine (adrenaline), and the skin pigment melanin.

   a. 1,2-Dioxetanedione
   b. Estradiol 17 beta-cypionate
   c. Phenylalanine
   d. Etilevodopa

ANSWER KEY
Chapter 7. Biological macromolecules: providing life's infrastructure

1. c
2. c
3. b
4. a
5. c

You can take the complete Chapter Practice Test

for Chapter 7. Biological macromolecules: providing life's infrastructure
on all key terms, persons, places, and concepts.

Online 99 Cents

http://www.epub14.51.19910.7.cram101.com/

Use www.Cram101.com for all your study needs

including Cram101's online interactive problem solving labs in chemistry, statistics, mathematics, and more.

CHAPTER OUTLINE: KEY TERMS, PEOPLE, PLACES, CONCEPTS
Chapter 8
Molecular shape and structure 1: from atoms to small molecules

_____ Bond length

_____ Atomic radius

_____ VSEPR theory

_____ Geometry

_____ Molecular geometry

_____ Racemic mixture

_____ Lewis structure

_____ DNA profiling

_____ Double bond

_____ Single bond

_____ Triple bond

_____ Valence electron

_____ Atomic orbital

_____ DNA-binding protein

_____ Molecule

_____ Rotation

_____ Alkane

_____ Dihedral angle

CHAPTER HIGHLIGHTS: KEY TERMS, PEOPLE, PLACES, CONCEPTS
Chapter 8. Molecular shape and structure 1: from atoms to small molecules

Bond length	In molecular geometry, bond length is the average distance between nuclei of two bonded atoms in a molecule. Explanation Bond length is related to bond order, when more electrons participate in bond formation the bond will get shorter. Bond length is also inversely related to bond strength and the bond dissociation energy, as (all other things being equal) a stronger bond will be shorter.
Atomic radius	The atomic radius of a chemical element is a measure of the size of its atoms, usually the mean or typical distance from the nucleus to the boundary of the surrounding cloud of electrons. Since the boundary is not a well-defined physical entity, there are various non-equivalent definitions of atomic radius. Depending on the definition, the term may apply only to isolated atoms, or also to atoms in condensed matter, covalently bound in molecules, or in ionized and excited states; and its value may be obtained through experimental measurements, or computed from theoretical models.
VSEPR theory	Valence shell electron pair repulsion (VSEPR) rules are a model in chemistry used to predict the shape of individual molecules based upon the extent of electron-pair electrostatic repulsion. It is also named Gillespie-Nyholm theory after its two main developers. The premise of VSEPR is that the valence electron pairs surrounding an atom mutually repel each other, and will therefore adopt an arrangement that minimizes this repulsion, thus determining the molecular geometry. The number of electron pairs surrounding an atom, both bonding and nonbonding, is called its steric number. VSEPR theory is usually compared and contrasted with valence bond theory, which addresses molecular shape through orbitals that are energetically accessible for bonding.

Chapter 8. Molecular shape and structure 1: from atoms to small molecules

Geometry	Geometry is a branch of mathematics concerned with questions of shape, size, relative position of figures, and the properties of space. A mathematician who works in the field of geometry is called a geometer. Geometry arose independently in a number of early cultures as a body of practical knowledge concerning lengths, areas, and volumes, with elements of a formal mathematical science emerging in the West as early as Thales (6th Century BC).
Molecular geometry	Molecular geometry is the three-dimensional arrangement of the atoms that constitute a molecule. It determines several properties of a substance including its reactivity, polarity, phase of matter, color, magnetism, and biological activity. The molecular geometry can be determined by various spectroscopic methods and diffraction methods.
Racemic mixture	In chemistry, a racemic mixture, is one that has equal amounts of left- and right-handed enantiomers of a chiral molecule. The first known racemic mixture was 'racemic acid', which Louis Pasteur found to be a mixture of the two enantiomeric isomers of tartaric acid. Nomenclature A racemic mixture is denoted by the prefix (±)- or dl- (for sugars the prefix DL- may be used), indicating an equal (1:1) mixture of dextro and levo isomers.
Lewis structure	Lewis structures (also known as Lewis dot diagrams, electron dot diagrams, and electron dot structures) are diagrams that show the bonding between atoms of a molecule and the lone pairs of electrons that may exist in the molecule.
DNA profiling	DNA profiling is a technique employed by forensic scientists to assist in the identification of individuals by their respective DNA profiles. DNA profiles are encrypted sets of numbers that reflect a person's DNA makeup, which can also be used as the person's identifier. DNA profiling should not be confused with full genome sequencing.
Double bond	A double bond in chemistry is a chemical bond between two chemical elements involving four bonding electrons instead of the usual two. The most common double bond, that between two carbon atoms, can be found in alkenes. Many types of double bonds between two different elements exist, for example in a carbonyl group with a carbon atom and an oxygen atom.

Chapter 8. Molecular shape and structure 1: from atoms to small molecules

Single bond	A Single bond in chemistry is a chemical bond between two chemical elements involving two bonding electrons. Usually, Single bond is Sigma bond. but diboron is Pi bond.(Molecular orbital diagram#Diboron MO diagram.
Triple bond	A triple bond in chemistry is a chemical bond between two chemical elements involving six bonding electrons instead of the usual two in a covalent single bond. The most common triple bond, that between two carbon atoms, can be found in alkynes. Other functional groups containing a triple bond are cyanides and isocyanides.
Valence electron	In chemistry, valence electrons are the electrons of an atom that can participate in the formation of chemical bonds with other atoms. Valence electrons are their 'own' electrons, present in the free neutral atom, that combine with valence electrons of other atoms to form chemical bonds. In a single covalent bond both atoms contribute one valence electron to form a shared pair.
Atomic orbital	An atomic orbital is a mathematical function that describes the wave-like behavior of either one electron or a pair of electrons in an atom. This function can be used to calculate the probability of finding any electron of an atom in any specific region around the atom's nucleus. The term may also refer to the physical region where the electron can be calculated to be, as defined by the particular mathematical form of the orbital.
DNA-binding protein	DNA-binding proteins are proteins that are composed of DNA-binding domains and thus have a specific or general affinity for either single or double stranded DNA. Sequence-specific DNA-binding proteins generally interact with the major groove of B-DNA, because it exposes more functional groups that identify a base pair. However there are some known minor groove DNA-binding ligands such as Netropsin, Distamycin, Hoechst 33258, Pentamidine and others. Examples DNA-binding proteins include transcription factors which modulate the process of transcription, various polymerases, nucleases which cleave DNA molecules, and histones which are involved in chromosome packaging and transcription in the cell nucleus.

Chapter 8. Molecular shape and structure 1: from atoms to small molecules

Molecule	A molecule is an electrically neutral group of two or more atoms held together by covalent chemical bonds. Molecules are distinguished from ions by their lack of electrical charge. However, in quantum physics, organic chemistry, and biochemistry, the term molecule is often used less strictly, also being applied to polyatomic ions.
Rotation	A rotation is a circular movement of an object around a center of rotation. A three-dimensional object rotates always around an imaginary line called a rotation axis. If the axis is within the body, and passes through its center of mass the body is said to rotate upon itself, or spin.
Alkane	Alkanes (also known as paraffins or saturated hydrocarbons) are chemical compounds that consist only of hydrogen and carbon atoms and are bonded exclusively by single bonds (i.e., they are saturated compounds) without any cycles . Alkanes belong to a homologous series of organic compounds in which the members differ by a constant relative molecular mass of 14. They have 2 main commercial sources, crude oil and natural gas. Each carbon atom has 4 bonds (either C-H or C-C bonds), and each hydrogen atom is joined to a carbon atom (H-C bonds).
Dihedral angle	In geometry, a dihedral or torsion angle is the angle between two planes. The dihedral angle of two planes can be seen by looking at the planes 'edge on', i.e., along their line of intersection. The dihedral angle φ_{AB} between two planes denoted A and B is the angle between their two normal unit vectors $\mathbf{n}_A$ and $\mathbf{n}_B$: $$\cos\varphi_{AB} = \mathbf{n}_A \cdot \mathbf{n}_B.$$ A dihedral angle can be signed; for example, the dihedral angle φ_{AB} can be defined as the angle through which plane A must be rotated (about their common line of intersection) to align it with plane B. Thus, $\varphi_{AB} = -\varphi_{BA}$.

PRACTICE QUIZ
Chapter 8. Molecular shape and structure 1: from atoms to small molecules

1. _____s (also known as Lewis dot diagrams, electron dot diagrams, and electron dot structures) are diagrams that show the bonding between atoms of a molecule and the lone pairs of electrons that may exist in the molecule.

    a. Ligand
    b. Ligand dependent pathway
    c. Lewis structure
    d. Ligand isomerism

2. In molecular geometry, _____ is the average distance between nuclei of two bonded atoms in a molecule.

    Explanation
    _____ is related to bond order, when more electrons participate in bond formation the bond will get shorter. _____ is also inversely related to bond strength and the bond dissociation energy, as (all other things being equal) a stronger bond will be shorter.

    a. Capped square antiprismatic molecular geometry
    b. Chicken wire
    c. Bond length
    d. LCP theory

3. The _____ of a chemical element is a measure of the size of its atoms, usually the mean or typical distance from the nucleus to the boundary of the surrounding cloud of electrons. Since the boundary is not a well-defined physical entity, there are various non-equivalent definitions of _____.

    Depending on the definition, the term may apply only to isolated atoms, or also to atoms in condensed matter, covalently bound in molecules, or in ionized and excited states; and its value may be obtained through experimental measurements, or computed from theoretical models.

    a. Ionic radius
    b. Atomic radius
    c. Cyclic compound
    d. LCP theory

---

4. In chemistry, a _____, is one that has equal amounts of left- and right-handed enantiomers of a chiral molecule. The first known _____ was 'racemic acid', which Louis Pasteur found to be a mixture of the two enantiomeric isomers of tartaric acid.

    Nomenclature
    A _____ is denoted by the prefix (±)- or dl- (for sugars the prefix DL- may be used), indicating an equal (1:1) mixture of dextro and levo isomers.

    a. Racemization
    b. Racemic mixture
    c. Regioselectivity
    d. Ring flip

5. A _____ is an electrically neutral group of two or more atoms held together by covalent chemical bonds. _____s are distinguished from ions by their lack of electrical charge. However, in quantum physics, organic chemistry, and biochemistry, the term _____ is often used less strictly, also being applied to polyatomic ions.

    a. Molecule
    b. Biomolecule
    c. Diatomic carbon
    d. Diatomic molecule

ANSWER KEY
Chapter 8. Molecular shape and structure 1: from atoms to small molecules

1. c
2. c
3. b
4. b
5. a

You can take the complete Chapter Practice Test

for Chapter 8. Molecular shape and structure 1: from atoms to small molecules
on all key terms, persons, places, and concepts.

Online 99 Cents

http://www.epub14.51.19910.8.cram101.com/

Use www.Cram101.com for all your study needs

including Cram101's online interactive problem solving labs in chemistry, statistics, mathematics, and more.

CHAPTER OUTLINE: KEY TERMS, PEOPLE, PLACES, CONCEPTS
Chapter 9
Molecular shape and structure 2: the shape of large molecules

- Avogadro constant
- Monomer
- Polymer
- VSEPR theory
- Molecular geometry
- Amylase
- Amylopectin
- Amylose
- Glycosidic bond
- DNA profiling
- DNA-binding protein
- Blood sugar
- Glucagon
- Aufbau principle
- Gel electrophoresis
- Nucleic acid
- Amino acid
- Electron pair
- Double Helix

Chapter 9. Molecular shape and structure 2: the shape of large molecules

- _____ Disulfide bond
- _____ ATP synthase
- _____ X-ray crystallography
- _____ Acetic acid
- _____ Chromatin
- _____ Euchromatin
- _____ Gene expression
- _____ Heterochromatin
- _____ Histone
- _____ Nucleosome
- _____ Phosphorylation
- _____ Enzyme
- _____ Transition state
- _____ Ubiquitin

CHAPTER HIGHLIGHTS: KEY TERMS, PEOPLE, PLACES, CONCEPTS
Chapter 9. Molecular shape and structure 2: the shape of large molecules

Avogadro constant	In chemistry and physics, the Avogadro constant is defined as the ratio of the number of constituent particles N (usually atoms or molecules) in a sample to the amount of substance n (unit mole) through the relationship $N_A = Nn$. Thus, it is the proportionality factor that relates the molar mass of an entity, i.e., the mass per amount of substance, to the mass of said entity. The Avogadro constant expresses the number of elementary entities per mole of substance and it has the value $6.022\ 141\ 29(27) \times 10^{23}\ mol^{-1}$.
Monomer	A monomer is a molecule that may bind chemically to other molecules to form a polymer. The term 'monomeric protein' may also be used to describe one of the proteins making up a multiprotein complex. The most common natural monomer is glucose, which is linked by glycosidic bonds into polymers such as cellulose and starch, and is over 77% of the mass of all plant matter.
Polymer	A polymer is a large molecule (macromolecule) composed of repeating structural units. These sub-units are typically connected by covalent chemical bonds. Although the term polymer is sometimes taken to refer to plastics, it actually encompasses a large class of compounds comprising both natural and synthetic materials with a wide variety of properties.
VSEPR theory	Valence shell electron pair repulsion (VSEPR) rules are a model in chemistry used to predict the shape of individual molecules based upon the extent of electron-pair electrostatic repulsion. It is also named Gillespie-Nyholm theory after its two main developers. The premise of VSEPR is that the valence electron pairs surrounding an atom mutually repel each other, and will therefore adopt an arrangement that minimizes this repulsion, thus determining the molecular geometry. The number of electron pairs surrounding an atom, both bonding and nonbonding, is called its steric number. VSEPR theory is usually compared and contrasted with valence bond theory, which addresses molecular shape through orbitals that are energetically accessible for bonding.
Molecular geometry	Molecular geometry is the three-dimensional arrangement of the atoms that constitute a molecule. It determines several properties of a substance including its reactivity, polarity, phase of matter, color, magnetism, and biological activity.

Chapter 9. Molecular shape and structure 2: the shape of large molecules

	The molecular geometry can be determined by various spectroscopic methods and diffraction methods.
Amylase	Amylase is an enzyme that catalyses the breakdown of starch into sugars. Amylase is present in human saliva, where it begins the chemical process of digestion. Foods that contain much starch but little sugar, such as rice and potato, taste slightly sweet as they are chewed because amylase turns some of their starch into sugar in the mouth.
Amylopectin	Amylopectin is a soluble polysaccharide and highly branched polymer of glucose found in plants. It is one of the two components of starch, the other being amylose. Glucose units are linked in a linear way with α(1→4) glycosidic bonds.
Amylose	Amylose is a linear polymer made up of D-glucose units. This polysaccharide is one of the two components of starch, making up approximately 20-30% of the structure. The other component is amylopectin, which makes up 70-80% of the structure.
Glycosidic bond	In chemistry, a glycosidic bond is a type of covalent bond that joins a carbohydrate (sugar) molecule to another group, which may or may not be another carbohydrate. A glycosidic bond is formed between the hemiacetal group of a saccharide and the hydroxyl group of some organic compound such as an alcohol. If the group attached to the carbohydrate residue is not another saccharide it is referred to as an aglycone.
DNA profiling	DNA profiling is a technique employed by forensic scientists to assist in the identification of individuals by their respective DNA profiles. DNA profiles are encrypted sets of numbers that reflect a person's DNA makeup, which can also be used as the person's identifier. DNA profiling should not be confused with full genome sequencing.

Chapter 9. Molecular shape and structure 2: the shape of large molecules

DNA-binding protein	DNA-binding proteins are proteins that are composed of DNA-binding domains and thus have a specific or general affinity for either single or double stranded DNA. Sequence-specific DNA-binding proteins generally interact with the major groove of B-DNA, because it exposes more functional groups that identify a base pair. However there are some known minor groove DNA-binding ligands such as Netropsin, Distamycin, Hoechst 33258, Pentamidine and others. Examples DNA-binding proteins include transcription factors which modulate the process of transcription, various polymerases, nucleases which cleave DNA molecules, and histones which are involved in chromosome packaging and transcription in the cell nucleus.
Blood sugar	The blood sugar concentration or blood glucose level is the amount of glucose (sugar) present in the blood of a human or animal. Normally in mammals, the body maintains the blood glucose level at a reference range between about 3.6 and 5.8 mM (mmol/L, i.e., millimoles/liter), or 64.8 and 104.4 mg/dL. The human body naturally tightly regulates blood glucose levels as a part of metabolic homeostasis. Glucose is the primary source of energy for the body's cells, and blood lipids (in the form of fats and oils) are primarily a compact energy store.
Glucagon	Glucagon, a peptide hormone secreted by the pancreas, raises blood glucose levels. Its effect is opposite that of insulin, which lowers blood glucose levels. The pancreas releases glucagon when blood sugar (glucose) levels fall too low.
Aufbau principle	The Aufbau principle is used to determine the electron configuration of an atom, molecule or ion. The principle postulates a hypothetical process in which an atom is 'built up' by progressively adding electrons. As they are added, they assume their most stable conditions (electron orbitals) with respect to the nucleus and those electrons already there.

Chapter 9. Molecular shape and structure 2: the shape of large molecules

Gel electrophoresis	Gel electrophoresis is a method used in clinical chemistry to separate proteins by charge and or size (IEF agarose, essentially size independent) and in biochemistry and molecular biology to separate a mixed population of DNA and RNA fragments by length, to estimate the size of DNA and RNA fragments or to separate proteins by charge. Nucleic acid molecules are separated by applying an electric field to move the negatively charged molecules through an agarose matrix. Shorter molecules move faster and migrate farther than longer ones because shorter molecules migrate more easily through the pores of the gel.
Nucleic acid	Nucleic acids are biological molecules essential for known forms of life on this planet; they include DNA (deoxyribonucleic acid) and RNA (ribonucleic acid). Together with proteins, nucleic acids are the most important biological macromolecules; each is found in abundance in all living things, where they function in encoding, transmitting and expressing genetic information.
	Nucleic acids were discovered by Friedrich Miescher in 1869. Experimental studies of nucleic acids constitute a major part of modern biological and medical research, and form a foundation for genome and forensic science, as well as the biotechnology and pharmaceutical industries.
Amino acid	Amino acids (?'mi?no?, ?'ma?o?, or 'æm?o?) are molecules containing an amine group, a carboxylic acid group, and a side-chain that is specific to each amino acid. The key elements of an amino acid are carbon, hydrogen, oxygen, and nitrogen. They are particularly important in biochemistry, where the term usually refers to alpha-amino acids.
Electron pair	In chemistry, an electron pair consists of two electrons that occupy the same orbital but have opposite spins.
	Because electrons are fermions, the Pauli exclusion principle forbids these particles from having exactly the same quantum numbers. Therefore the only way to occupy the same orbital, i.e. have the same orbital quantum numbers, is to differ in the spin quantum number.
Double Helix	Double Helix (2004), a novel by Nancy Werlin, is about 18-year old Eli Samuels, who works for a famous molecular biologist named Dr. Quincy Wyatt. There is a mysterious connection between Dr. Wyatt and Eli's parents, and all Eli knows about the connection is that it has something to do with his mother, who has Huntington's disease. Because of the connection between Dr. Wyatt and the Samuels family, Eli's father is strongly against Eli working there.

Chapter 9. Molecular shape and structure 2: the shape of large molecules

Disulfide bond	In chemistry, a disulfide bond is a covalent bond, usually derived by the coupling of two thiol groups. The linkage is also called an SS-bond or disulfide bridge. The overall connectivity is therefore R-S-S-R. The terminology is widely used in biochemistry.
ATP synthase	ATP synthase is an important enzyme that provides energy for the cell to use through the synthesis of adenosine triphosphate (ATP). ATP is the most commonly used 'energy currency' of cells from most organisms. It is formed from adenosine diphosphate (ADP) and inorganic phosphate (P_i), and needs energy.
X-ray crystallography	X-ray crystallography is a method of determining the arrangement of atoms within a crystal, in which a beam of X-rays strikes a crystal and causes the beam of light to spread into many specific directions. From the angles and intensities of these diffracted beams, a crystallographer can produce a three-dimensional picture of the density of electrons within the crystal. From this electron density, the mean positions of the atoms in the crystal can be determined, as well as their chemical bonds, their disorder and various other information.
Acetic acid	Acetic acid ?'si?t?k is an organic compound with the chemical formula CH_3CO_2H (also written as CH_3COOH). It is a colourless liquid that when undiluted is also called glacial acetic acid. Acetic acid is the main component of vinegar (apart from water), and has a distinctive sour taste and pungent smell.
Chromatin	Chromatin is the combination of DNA and proteins that make up the contents of the nucleus of a cell. The primary functions of chromatin are: to package DNA into a smaller volume to fit in the cell, to strengthen the DNA to allow mitosis and meiosis and prevent DNA damage, and to control gene expression and DNA replication. The primary protein components of chromatin are histones that compact the DNA. Chromatin is only found in eukaryotic cells: prokaryotic cells have a very different organization of their DNA which is referred to as a genophore (a chromosome without chromatin).
Euchromatin	Euchromatin is a lightly packed form of chromatin (DNA, RNA and protein) that is rich in gene concentration, and is often (but not always) under active transcription. Unlike heterochromatin, it is found in both cells with nuclei (eukaryotes) and cells without nuclei (prokaryotes). Euchromatin comprises the most active portion of the genome within the cell nucleus.

Chapter 9. Molecular shape and structure 2: the shape of large molecules

Gene expression	Gene expression is the process by which information from a gene is used in the synthesis of a functional gene product. These products are often proteins, but in non-protein coding genes such as ribosomal RNA (rRNA), transfer RNA (tRNA) or small nuclear RNA (snRNA) genes, the product is a functional RNA. The process of gene expression is used by all known life - eukaryotes (including multicellular organisms), prokaryotes (bacteria and archaea), possibly induced by viruses - to generate the macromolecular machinery for life. Several steps in the gene expression process may be modulated, including the transcription, RNA splicing, translation, and post-translational modification of a protein.
Heterochromatin	Heterochromatin is a tightly packed form of DNA, which comes in different varieties. These varieties lie on a continuum between the two extremes of constitutive and facultative heterochromatin. Both play a role in the expression of genes, where constitutive heterochromatin can affect the genes near them (position-effect variegation) and where facultative heterochromatin is the result of genes that are silenced through a mechanism such as histone methylation or siRNA through RNAi.
Histone	In biology, histones are highly alkaline proteins found in eukaryotic cell nuclei that package and order the DNA into structural units called nucleosomes. They are the chief protein components of chromatin, acting as spools around which DNA winds, and play a role in gene regulation. Without histones, the unwound DNA in chromosomes would be very long (a length to width ratio of more than 10 million to one in human DNA).
Nucleosome	Nucleosomes are the basic unit of DNA packaging in eukaryotes, consisting of a segment of DNA wound around a histone protein core. This structure is often compared to thread wrapped around a spool. Nucleosomes form the fundamental repeating units of eukaryotic chromatin, which is used to pack the large eukaryotic genomes into the nucleus while still ensuring appropriate access to it (in mammalian cells approximately 2 m of linear DNA have to be packed into a nucleus of roughly 10 μm diameter). Nucleosomes are folded through a series of successively higher order structures to eventually form a chromosome; this both compacts DNA and creates an added layer of regulatory control which ensures correct gene expression. Nucleosomes are thought to carry epigenetically inherited information in the form of covalent modifications of their core histones.

Chapter 9. Molecular shape and structure 2: the shape of large molecules

Phosphorylation	Phosphorylation is the addition of a phosphate (PO_4^{3-}) group to a protein or other organic molecule. Phosphorylation turns many protein enzymes on and off, thereby altering their function and activity.
	Protein phosphorylation in particular plays a significant role in a wide range of cellular processes.
Enzyme	Enzymes ( ) are biological molecules that catalyze (i.e., increase the rates of) chemical reactions. In enzymatic reactions, the molecules at the beginning of the process, called substrates, are converted into different molecules, called products. Almost all chemical reactions in a biological cell need enzymes in order to occur at rates sufficient for life.
Transition state	The transition state of a chemical reaction is a particular configuration along the reaction coordinate. It is defined as the state corresponding to the highest energy along this reaction coordinate. At this point, assuming a perfectly irreversible reaction, colliding reactant molecules will always go on to form products.

Ubiquitin family

Ubiquitin	Ubiquitin is a small regulatory protein that has been found in almost all tissues (ubiquitously) of eukaryotic organisms. Among other functions, it directs protein recycling. Ubiquitin can be attached to proteins and label them for destruction.

PRACTICE QUIZ
Chapter 9. Molecular shape and structure 2: the shape of large molecules

1. The _____ is used to determine the electron configuration of an atom, molecule or ion. The principle postulates a hypothetical process in which an atom is 'built up' by progressively adding electrons. As they are added, they assume their most stable conditions (electron orbitals) with respect to the nucleus and those electrons already there.

    a. Autoionization
    b. Azimuthal quantum number
    c. Aufbau principle
    d. Electric field gradient

2. In biology, _____s are highly alkaline proteins found in eukaryotic cell nuclei that package and order the DNA into structural units called nucleosomes. They are the chief protein components of chromatin, acting as spools around which DNA winds, and play a role in gene regulation. Without _____s, the unwound DNA in chromosomes would be very long (a length to width ratio of more than 10 million to one in human DNA).

    a. Homeotic gene
    b. Histone
    c. Homogeneously staining region
    d. HomoloGene

3. _____ is an important enzyme that provides energy for the cell to use through the synthesis of adenosine triphosphate (ATP). ATP is the most commonly used 'energy currency' of cells from most organisms. It is formed from adenosine diphosphate (ADP) and inorganic phosphate (P_i), and needs energy.

    a. Ecarin
    b. Endoenzyme
    c. ATP synthase
    d. Endoglycosidase H

4. _____ is an enzyme that catalyses the breakdown of starch into sugars. _____ is present in human saliva, where it begins the chemical process of digestion. Foods that contain much starch but little sugar, such as rice and potato, taste slightly sweet as they are chewed because _____ turns some of their starch into sugar in the mouth.

    a. Amylase
    b. Arachidonate 5-lipoxygenase
    c. Argininosuccinate synthase
    d. Artificial enzyme

---

5. In chemistry and physics, the _____ is defined as the ratio of the number of constituent particles N (usually atoms or molecules) in a sample to the amount of substance n (unit mole) through the relationship $N_A = Nn$. Thus, it is the proportionality factor that relates the molar mass of an entity, i.e., the mass per amount of substance, to the mass of said entity. The _____ expresses the number of elementary entities per mole of substance and it has the value $6.022\ 141\ 29(27) \times 10^{23}\ mol^{-1}$.

   a. Equivalent weight
   b. Avogadro constant
   c. Osmometer
   d. Osmotic pressure

ANSWER KEY
Chapter 9. Molecular shape and structure 2: the shape of large molecules

1. c
2. b
3. c
4. a
5. b

You can take the complete Chapter Practice Test

for Chapter 9. Molecular shape and structure 2: the shape of large molecules
on all key terms, persons, places, and concepts.

Online 99 Cents

http://www.epub14.51.19910.9.cram101.com/

Use www.Cram101.com for all your study needs

including Cram101's online interactive problem solving labs in chemistry, statistics, mathematics, and more.

CHAPTER OUTLINE: KEY TERMS, PEOPLE, PLACES, CONCEPTS
Chapter 10
Isomerism: generating chemical variety

- Intramolecular force
- DNA profiling
- Graphene
- Structural formula
- Alkene
- Cistrans isomerism
- Enantiomer
- Dichloromethane
- Optical rotation
- X-ray crystallography
- Amino acid
- Avogadro constant
- Lewis structure
- Cell wall
- Vancomycin
- Fumaric acid
- Maleic acid
- Heroin
- Medicinal chemistry

Chapter 10. Isomerism: generating chemical variety

- Morphine
- Racemic mixture

CHAPTER HIGHLIGHTS: KEY TERMS, PEOPLE, PLACES, CONCEPTS
Chapter 10. Isomerism: generating chemical variety

Intramolecular force	An intramolecular force is any force that holds together the atoms making up a molecule or compound. They contain all types of chemical bond. They are stronger than intermolecular forces, which are present between atoms or molecules that are not actually bonded.
DNA profiling	DNA profiling is a technique employed by forensic scientists to assist in the identification of individuals by their respective DNA profiles. DNA profiles are encrypted sets of numbers that reflect a person's DNA makeup, which can also be used as the person's identifier. DNA profiling should not be confused with full genome sequencing.
Graphene	Graphene is an allotrope of carbon. Its structure is one-atom-thick planar sheets of sp^2-bonded carbon atoms that are densely packed in a honeycomb crystal lattice. The term graphene was coined as a combination of graphite and the suffix -ene by Hanns-Peter Boehm, who described single-layer carbon foils in 1962. Graphene is most easily visualized as an atomic-scale chicken wire made of carbon atoms and their bonds.
Structural formula	The structural formula of a chemical compound is a graphical representation of the molecular structure, showing how the atoms are arranged. The chemical bonding within the molecule is also shown, either explicitly or implicitly. Also several other formats are used, as in chemical databases, such as SMILES, InChI and CML. Unlike chemical formulas or chemical names, structural formulas provide a representation of the molecular structure.
Alkene	In organic chemistry, an alkene, olefin, or olefine is an unsaturated chemical compound containing at least one carbon-to-carbon double bond. The simplest acyclic alkenes, with only one double bond and no other functional groups, form an homologous series of hydrocarbons with the general formula C_nH_{2n}. The simplest alkene is ethylene (C_2H_4), which has the International Union of Pure and Applied Chemistry (IUPAC) name ethene.

Chapter 10. Isomerism: generating chemical variety

Cistrans isomerism	In organic chemistry, cistrans isomerism (also known as geometric isomerism, configuration isomerism, or EZ isomerism) is a form of stereoisomerism describing the orientation of functional groups within a molecule. In general, such isomers contain double bonds, which cannot rotate, but they can also arise from ring structures, wherein the rotation of bonds is greatly restricted. Cis and trans isomers occur both in organic molecules and in inorganic coordination complexes.
Enantiomer	In chemistry, an enantiomer is one of two stereoisomers that are mirror images of each other that are non-superposable (not identical), much as one's left and right hands are the same except for opposite orientation.
	Organic compounds that contain an asymmetric (chiral) Carbon usually have two non-superimposable structures. These two structures are mirror images of each other and are, thus, commonly called enantiomorphs (enantio = opposite ; morph = form) Hence, optical isomerism (which occurs due to these same mirror-image properties) is now commonly referred to as enantiomerism
	Enantiopure compounds refer to samples having, within the limits of detection, molecules of only one chirality.
Dichloromethane	Dichloromethane -- or methylene chloride -- is an organic compound with the formula CH_2Cl_2. This colorless, volatile liquid with a moderately sweet aroma is widely used as a solvent. Although it is not miscible with water, it is miscible with many organic solvents.
Optical rotation	Optical rotation is the turning of the plane of linearly polarized light about the direction of motion as the light travels through certain materials. It occurs in solutions of chiral molecules such as sucrose (sugar), solids with rotated crystal planes such as quartz, and spin-polarized gases of atoms or molecules. It is used in the sugar industry to measure syrup concentration, in optics to manipulate polarization, in chemistry to characterize substances in solution, and in optical mineralogy to help identify certain minerals in thin sections.

Chapter 10. Isomerism: generating chemical variety

X-ray crystallography	X-ray crystallography is a method of determining the arrangement of atoms within a crystal, in which a beam of X-rays strikes a crystal and causes the beam of light to spread into many specific directions. From the angles and intensities of these diffracted beams, a crystallographer can produce a three-dimensional picture of the density of electrons within the crystal. From this electron density, the mean positions of the atoms in the crystal can be determined, as well as their chemical bonds, their disorder and various other information.
Amino acid	Amino acids (?'mi?no?, ?'ma?o?, or 'æm?o?) are molecules containing an amine group, a carboxylic acid group, and a side-chain that is specific to each amino acid. The key elements of an amino acid are carbon, hydrogen, oxygen, and nitrogen. They are particularly important in biochemistry, where the term usually refers to alpha-amino acids.
Avogadro constant	In chemistry and physics, the Avogadro constant is defined as the ratio of the number of constituent particles N (usually atoms or molecules) in a sample to the amount of substance n (unit mole) through the relationship $N_A = Nn$. Thus, it is the proportionality factor that relates the molar mass of an entity, i.e., the mass per amount of substance, to the mass of said entity. The Avogadro constant expresses the number of elementary entities per mole of substance and it has the value $6.022\ 141\ 29(27) \times 10^{23}\ mol^{-1}$.
Lewis structure	Lewis structures (also known as Lewis dot diagrams, electron dot diagrams, and electron dot structures) are diagrams that show the bonding between atoms of a molecule and the lone pairs of electrons that may exist in the molecule.
Cell wall	The cell wall is the tough, usually flexible but sometimes fairly rigid layer that surrounds some types of cells. It is located outside the cell membrane and provides these cells with structural support and protection, in addition to acting as a filtering mechanism. A major function of the cell wall is to act as a pressure vessel, preventing over-expansion when water enters the cell.
Vancomycin	Vancomycin INN ( ) is a glycopeptide antibiotic used in the prophylaxis and treatment of infections caused by Gram-positive bacteria. Vancomycin was first isolated in 1953 at Eli Lilly, from a soil sample collected from the interior jungles of Borneo by a missionary. It is a naturally occuring antibiotic made by the soil bacterium Actinobacteria species Amycolatopsis orientalis (formerly designated Nocardia orientalis).
Fumaric acid	Fumaric acid is the chemical compound with the formula $HO_2CCH=CHCO_2H$. This white crystalline compound is one of two isomeric unsaturated dicarboxylic acids, the other being maleic acid. In fumaric acid the carboxylic acid groups are trans (E) and in maleic acid they are cis (Z). Fumaric acid has a fruit-like taste.

Chapter 10. Isomerism: generating chemical variety

Maleic acid	Maleic acid is an organic compound that is a dicarboxylic acid, a molecule with two carboxyl groups. Maleic acid is the cis-isomer of butenedioic acid, whereas fumaric acid is the trans-isomer. It is mainly used as a precursor to fumaric acid, and relative to its parent maleic anhydride, maleic acid has few applications.
Heroin	Heroin (diacetylmorphine (INN)), also known as diamorphine (BAN), is a semi-synthetic opioid drug synthesized from morphine, a derivative of the opium poppy. It is the 3,6-diacetyl ester of morphine (di (two)-acetyl-morphine). The white crystalline form is commonly the hydrochloride salt diacetylmorphine hydrochloride, though often adulterated thus dulling the sheen and consistency from that to a matte white powder, which diacetylmorphine freebase typically is. 90% of diacetylmorphine is thought to be produced in Afghanistan.
Medicinal chemistry	Medicinal chemistry and pharmaceutical chemistry are disciplines at the intersection of chemistry, especially synthetic organic chemistry, and pharmacology and various other biological specialties, where they are involved with design, chemical synthesis and development for market of pharmaceutical agents (drugs). Compounds used as medicines are most often organic compounds, which are often divided into the broad classes of small organic molecules (e.g., atorvastatin, fluticasone, clopidogrel) and 'biologics' (infliximab, erythropoietin, insulin glargine), the latter of which are most often medicinal preparations of proteins (natural and recombinant antibodies, hormones, etc).. Inorganic and organometallic compounds are also useful as drugs (e.g., lithium and platinum-based agents such as lithium carbonate and cis-platin.
Morphine	Morphine (; MS Contin, MSIR, Avinza, Kadian, Oramorph, Roxanol, Kapanol) is a potent opiate analgesic drug that is used to relieve severe pain. It was first isolated in 1804 by Friedrich Sertürner, first distributed by him in 1817, and first commercially sold by Merck in 1827, which at the time was a single small chemists' shop. It was more widely used after the invention of the hypodermic needle in 1857. It took its name from the Greek god of dreams Morpheus .
Racemic mixture	In chemistry, a racemic mixture, is one that has equal amounts of left- and right-handed enantiomers of a chiral molecule. The first known racemic mixture was 'racemic acid', which Louis Pasteur found to be a mixture of the two enantiomeric isomers of tartaric acid. Nomenclature

Chapter 10. Isomerism: generating chemical variety

A racemic mixture is denoted by the prefix (±)- or dl- (for sugars the prefix DL- may be used), indicating an equal (1:1) mixture of dextro and levo isomers.

PRACTICE QUIZ
Chapter 10. Isomerism: generating chemical variety

1. In chemistry and physics, the _____ is defined as the ratio of the number of constituent particles N (usually atoms or molecules) in a sample to the amount of substance n (unit mole) through the relationship $N_A = Nn$. Thus, it is the proportionality factor that relates the molar mass of an entity, i.e., the mass per amount of substance, to the mass of said entity. The _____ expresses the number of elementary entities per mole of substance and it has the value $6.022\ 141\ 29(27) \times 10^{23}\ mol^{-1}$.

    a. Equivalent weight
    b. Osmolarity
    c. Avogadro constant
    d. Osmotic pressure

2. _____ is a technique employed by forensic scientists to assist in the identification of individuals by their respective DNA profiles. DNA profiles are encrypted sets of numbers that reflect a person's DNA makeup, which can also be used as the person's identifier. _____ should not be confused with full genome sequencing.

    a. DNA profiling
    b. Peak calling
    c. DNA supercoil
    d. DNA-DNA hybridization

3. _____ is the chemical compound with the formula $HO_2CCH=CHCO_2H$. This white crystalline compound is one of two isomeric unsaturated dicarboxylic acids, the other being maleic acid. In _____ the carboxylic acid groups are trans (E) and in maleic acid they are cis (Z). _____ has a fruit-like taste.

    a. Fumaric acid
    b. Pyrocitric
    c. Succinic acid
    d. Crown sprouting

4. An _____ is any force that holds together the atoms making up a molecule or compound. They contain all types of chemical bond. They are stronger than intermolecular forces, which are present between atoms or molecules that are not actually bonded.

    a. Ionic bond
    b. Intramolecular force
    c. Octet rule
    d. Open shell

---

5. In organic chemistry, an _____, olefin, or olefine is an unsaturated chemical compound containing at least one carbon-to-carbon double bond. The simplest acyclic _____s, with only one double bond and no other functional groups, form an homologous series of hydrocarbons with the general formula C_nH_{2n}.

The simplest _____ is ethylene (C_2H_4), which has the International Union of Pure and Applied Chemistry (IUPAC) name ethene.

   a. Alkene
   b. Ethylenediamine pyrocatechol
   c. Explosophore
   d. In-Methylcyclophane

ANSWER KEY
Chapter 10. Isomerism: generating chemical variety

1. c
2. a
3. a
4. b
5. a

You can take the complete Chapter Practice Test

for Chapter 10. Isomerism: generating chemical variety
on all key terms, persons, places, and concepts.

Online 99 Cents

http://www.epub14.51.19910.10.cram101.com/

Use www.Cram101.com for all your study needs

including Cram101's online interactive problem solving labs in chemistry, statistics, mathematics, and more.

CHAPTER OUTLINE: KEY TERMS, PEOPLE, PLACES, CONCEPTS
Chapter 11
Chemical analysis 1: how do we know what is there?

- Blood sugar
- Gel electrophoresis
- Solvent
- Chemical species
- Chromatography
- Hydrophobicity
- Elution
- Racemic mixture
- Affinity
- Affinity chromatography
- Gas chromatography
- Metabolomics
- SDS-PAGE
- Acrylamide
- Agarose
- High-performance liquid chromatography
- DNA profiling
- Intramolecular force
- Isoelectric focusing

Chapter 11. Chemical analysis 1: how do we know what is there?

_____ Electron pair

_____ X-ray crystallography

_____ Centrifugation

_____ Gravitational field

_____ Sedimentation

_____ Mass spectrometry

_____ Avogadro constant

_____ Atomic mass

_____ Atomic mass unit

_____ Momentum

_____ Acceleration

_____ Chemical ionization

_____ Ionization

_____ Vaporization

_____ Electrospray ionization

_____ Fast atom bombardment

_____ Mass spectrum

_____ Molecular structure

_____ Coupling

Chapter 11. Chemical analysis 1: how do we know what is there?

- Gas chromatography-mass spectrometry
- Liquid chromatography-mass spectrometry
- Tandem mass spectrometry
- Spectroscopy
- Electromagnetic radiation
- Electromagnetic spectrum
- Rotation
- Vibration
- Transmission
- Transmittance
- Magnetic field
- Nuclear magnetic resonance
- Nuclear magnetic resonance spectroscopy
- Chemical shift
- Functional magnetic resonance imaging
- Magnetic resonance imaging
- Frequency
- Infrared spectroscopy
- Wavenumber

Chapter 11. Chemical analysis 1: how do we know what is there?

Diffraction

CHAPTER HIGHLIGHTS: KEY TERMS, PEOPLE, PLACES, CONCEPTS
Chapter 11. Chemical analysis 1: how do we know what is there?

Blood sugar	The blood sugar concentration or blood glucose level is the amount of glucose (sugar) present in the blood of a human or animal. Normally in mammals, the body maintains the blood glucose level at a reference range between about 3.6 and 5.8 mM (mmol/L, i.e., millimoles/liter), or 64.8 and 104.4 mg/dL. The human body naturally tightly regulates blood glucose levels as a part of metabolic homeostasis. Glucose is the primary source of energy for the body's cells, and blood lipids (in the form of fats and oils) are primarily a compact energy store.
Gel electrophoresis	Gel electrophoresis is a method used in clinical chemistry to separate proteins by charge and or size (IEF agarose, essentially size independent) and in biochemistry and molecular biology to separate a mixed population of DNA and RNA fragments by length, to estimate the size of DNA and RNA fragments or to separate proteins by charge. Nucleic acid molecules are separated by applying an electric field to move the negatively charged molecules through an agarose matrix. Shorter molecules move faster and migrate farther than longer ones because shorter molecules migrate more easily through the pores of the gel.
Solvent	A solvent is a liquid, solid, or gas that dissolves another solid, liquid, or gaseous solute, resulting in a solution that is soluble in a certain volume of solvent at a specified temperature. Common uses for organic solvents are in dry cleaning (e.g., tetrachloroethylene), as paint thinners (e.g., toluene, turpentine), as nail polish removers and glue solvents (acetone, methyl acetate, ethyl acetate), in spot removers (e.g., hexane, petrol ether), in detergents (citrus terpenes), in perfumes (ethanol), nail polish and in chemical synthesis. The use of inorganic solvents (other than water) is typically limited to research chemistry and some technological processes.
Chemical species	Chemical species are atoms, molecules, molecular fragments, ions, etc., subjected to a chemical process or to a measurement. Generally, a chemical species can be defined as an ensemble of chemically identical molecular entities that can explore the same set of molecular energy levels on a characteristic or delineated time scale. The term may be applied equally to a set of chemically identical atomic or molecular structural units in a solid array.
Chromatography	Chromatography is the collective term for a set of laboratory techniques for the separation of mixtures. The mixture is dissolved in a fluid called the mobile phase, which carries it through a structure holding another material called the stationary phase. The various constituents of the mixture travel at different speeds, causing them to separate.

Chapter 11. Chemical analysis 1: how do we know what is there?

Hydrophobicity	In chemistry, hydrophobicity is the physical property of a molecule (known as a hydrophobe) that is repelled from a mass of water. Hydrophobic molecules tend to be non-polar and thus prefer other neutral molecules and non-polar solvents. Hydrophobic molecules in water often cluster together forming micelles.
Elution	Elution is a term used in analytical and organic chemistry to describe the process of extracting one material from another by washing with a solvent (as in washing of loaded ion-exchange resins to remove captured ions). In a liquid chromatography experiment, for example, an analyte is generally adsorbed, or 'bound to', an adsorbent in a liquid chromatography column. The adsorbent, a solid phase (stationary phase), is a powder which is coated onto a solid support.
Racemic mixture	In chemistry, a racemic mixture, is one that has equal amounts of left- and right-handed enantiomers of a chiral molecule. The first known racemic mixture was 'racemic acid', which Louis Pasteur found to be a mixture of the two enantiomeric isomers of tartaric acid. Nomenclature A racemic mixture is denoted by the prefix (±)- or dl- (for sugars the prefix DL- may be used), indicating an equal (1:1) mixture of dextro and levo isomers.
Affinity	Affinity (taxonomy) - mainly in life sciences or natural history - refers to resemblance suggesting a common descent, phylogenetic relationship, or type. The term does, however, have broader application, such as in geology (for example, in descriptive and theoretical works), and similarly in astronomy. Basis In taxonomy the basis of any particular type of classification is the way in which objects in the domain resemble each other.

Chapter 11. Chemical analysis 1: how do we know what is there?

Affinity chromatography	Affinity chromatography is a method of separating biochemical mixtures and based on a highly specific interaction such as that between antigen and antibody, enzyme and substrate, or receptor and ligand. Uses Affinity chromatography can be used to: • Purify and concentrate a substance from a mixture into a buffering solution • Reduce the amount of a substance in a mixture • Discern what biological compounds bind to a particular substance • Purify and concentrate an enzyme solution. Principle The immobile phase is typically a gel matrix, often of agarose; a linear sugar molecule derived from algae. Usually the starting point is an undefined heterogeneous group of molecules in solution, such as a cell lysate, growth medium or blood serum.
Gas chromatography	Gas chromatography is a common type of chromatography used in analytical chemistry for separating and analyzing compounds that can be vaporized without decomposition. Typical uses of GC include testing the purity of a particular substance, or separating the different components of a mixture (the relative amounts of such components can also be determined). In some situations, GC may help in identifying a compound.
Metabolomics	Metabolomics is the scientific study of chemical processes involving metabolites. Specifically, metabolomics is the 'systematic study of the unique chemical fingerprints that specific cellular processes leave behind', the study of their small-molecule metabolite profiles. The metabolome represents the collection of all metabolites in a biological cell, tissue, organ or organism, which are the end products of cellular processes.
SDS-PAGE	SDS-PAGE, sodium dodecyl sulfate polyacrylamide gel electrophoresis, describes a technique widely used in biochemistry, forensics, genetics and molecular biology to separate proteins according to their electrophoretic mobility (a function of the length of a polypeptide chain and its charge) and no other physical feature. SDS is an anionic detergent applied to protein sample to linearize proteins and to impart a negative charge to linearized proteins. In most proteins, the binding of SDS to the polypeptide chain imparts an even distribution of charge per unit mass, thereby resulting in a fractionation by approximate size during electrophoresis.

Chapter 11. Chemical analysis 1: how do we know what is there?

Acrylamide	Acrylamide is a chemical compound with the chemical formula C_3H_5NO. Its IUPAC name is prop-2-enamide. It is a white odourless crystalline solid, soluble in water, ethanol, ether, and chloroform. Acrylamide is incompatible with acids, bases, oxidizing agents, iron, and iron salts.
Agarose	Agar or agar-agar is a gelatinous substance derived by boiling from a polysaccharide in red algae, where it accumulates in the cell walls of agarophyte and serve as the primary structural support for the algae's cell walls. Agar is a mixture of two components: the linear polysaccharide agarose, and a heterogeneous mixture of smaller molecules called agaropectin. Throughout history into modern times, agar has been chiefly used as an ingredient in desserts throughout Asia and also as a solid substrate to contain culture medium for microbiological work.
High-performance liquid chromatography	High-performance liquid chromatography HPLC, is a chromatographic technique used to separate a mixture of compounds in analytical chemistry and biochemistry with the purpose of identifying, quantifying and purifying the individual components of the mixture. Some common examples are the separation and quantitation of performance enhancement drugs (e.g. steroids) in urine samples, or of vitamin D levels in serum. HPLC typically utilizes different types of stationary phases (i.e. sorbents) contained in columns, a pump that moves the mobile phase and sample components through the column, and a detector capable of providing characteristic retention times for the sample components and area counts reflecting the amount of each analyte passing through the detector.
DNA profiling	DNA profiling is a technique employed by forensic scientists to assist in the identification of individuals by their respective DNA profiles. DNA profiles are encrypted sets of numbers that reflect a person's DNA makeup, which can also be used as the person's identifier. DNA profiling should not be confused with full genome sequencing.
Intramolecular force	An intramolecular force is any force that holds together the atoms making up a molecule or compound. They contain all types of chemical bond. They are stronger than intermolecular forces, which are present between atoms or molecules that are not actually bonded.

Chapter 11. Chemical analysis 1: how do we know what is there?

Isoelectric focusing	Isoelectric focusing also known as electrofocusing, is a technique for separating different molecules by their electric charge differences. It is a type of zone electrophoresis, usually performed on proteins in a gel, that takes advantage of the fact that overall charge on the molecule of interest is a function of the pH of its surroundings. Isoelectric focusing in laboratory IEF involves adding an ampholyte solution into immobilized pH gradient (IPG) gels.
Electron pair	In chemistry, an electron pair consists of two electrons that occupy the same orbital but have opposite spins. Because electrons are fermions, the Pauli exclusion principle forbids these particles from having exactly the same quantum numbers. Therefore the only way to occupy the same orbital, i.e. have the same orbital quantum numbers, is to differ in the spin quantum number.
X-ray crystallography	X-ray crystallography is a method of determining the arrangement of atoms within a crystal, in which a beam of X-rays strikes a crystal and causes the beam of light to spread into many specific directions. From the angles and intensities of these diffracted beams, a crystallographer can produce a three-dimensional picture of the density of electrons within the crystal. From this electron density, the mean positions of the atoms in the crystal can be determined, as well as their chemical bonds, their disorder and various other information.
Centrifugation	Centrifugation is a process that involves the use of the centrifugal force for the sedimentation of mixtures with a centrifuge, used in industry and in laboratory settings. More-dense components of the mixture migrate away from the axis of the centrifuge, while less-dense components of the mixture migrate towards the axis. Chemists and biologists may increase the effective gravitational force on a test tube so as to more rapidly and completely cause the precipitate ('pellet') to gather on the bottom of the tube.
Gravitational field	The gravitational field is a model used in physics to explain the existence of gravity. In its original concept, gravity was a force between point masses. Following Newton, Laplace attempted to model gravity as some kind of radiation field or fluid, and since the 19th century explanations for gravity have usually been sought in terms of a field model, rather than a point attraction.

Chapter 11. Chemical analysis 1: how do we know what is there?

Sedimentation	Sedimentation is the tendency for particles in suspension to settle out of the fluid in which they are entrained, and come to rest against a barrier. This is due to their motion through the fluid in response to the forces acting on them: these forces can be due to gravity, centrifugal acceleration or electromagnetism. In geology sedimentation is often used as the polar opposite of erosion, i.e., the terminal end of sediment transport.
Mass spectrometry	Mass spectrometry is an analytical technique that measures the mass-to-charge ratio of charged particles. It is used for determining masses of particles, for determining the elemental composition of a sample or molecule, and for elucidating the chemical structures of molecules, such as peptides and other chemical compounds. MS works by ionizing chemical compounds to generate charged molecules or molecule fragments and measuring their mass-to-charge ratios.
Avogadro constant	In chemistry and physics, the Avogadro constant is defined as the ratio of the number of constituent particles N (usually atoms or molecules) in a sample to the amount of substance n (unit mole) through the relationship $N_A = Nn$. Thus, it is the proportionality factor that relates the molar mass of an entity, i.e., the mass per amount of substance, to the mass of said entity. The Avogadro constant expresses the number of elementary entities per mole of substance and it has the value $6.022\ 141\ 29(27) \times 10^{23}\ mol^{-1}$.
Atomic mass	The atomic mass is the mass of a specific isotope, most often expressed in unified atomic mass units. The atomic mass is the total mass of protons, neutrons and electrons in a single atom. The atomic mass is sometimes incorrectly used as a synonym of relative atomic mass, average atomic mass and atomic weight; these differ subtly from the atomic mass.
Atomic mass unit	The unified atomic mass unit or dalton (symbol: Da) is a unit that is used for indicating mass on an atomic or molecular scale. It is defined as one twelfth of the rest mass of an unbound neutral atom of carbon-12 in its nuclear and electronic ground state, and has a value of $1.660\ 538\ 921(73) \times 10^{-27}$ kg. One dalton is approximately equal to the mass of one proton or one neutron; an equivalence of saying $1\ g\ mol^{-1}$.

Chapter 11. Chemical analysis 1: how do we know what is there?

Momentum	In classical mechanics, linear momentum or translational momentum is the product of the mass and velocity of an object: $$\mathbf{p} \equiv m\mathbf{v}.$$ Like velocity, linear momentum is a vector quantity, possessing a direction as well as a magnitude. Linear momentum is also a conserved quantity, meaning that if a closed system is not affected by external forces, its total linear momentum cannot change.
Acceleration	In physics, acceleration is the rate of change of velocity with time. In one dimension, acceleration is the rate at which something speeds up or slows down. For example, a car driving away (from standstill) is increasing its speed and is thus accelerating.
Chemical ionization	Chemical ionization is an ionization technique used in mass spectrometry. Chemical ionization is a lower energy process than electron ionization. The lower energy yields less fragmentation, and usually a simpler spectrum.
Ionization	Ionization is the process of converting an atom or molecule into an ion by adding or removing charged particles such as electrons or ions. In the case of ionisation of a gas, ion-pairs are created consisting of a free electron and a +ve ion. Types of Ionisation The process of ionization works slightly differently depending on whether an ion with a positive or a negative electric charge is being produced.
Vaporization	Vaporization of an element or compound is a phase transition from the liquid phase to gas phase. There are two types of vaporization: evaporation and boiling. Evaporation is a phase transition from the liquid phase to gas phase that occurs at temperatures below the boiling temperature at a given pressure.

Chapter 11. Chemical analysis 1: how do we know what is there?

Electrospray ionization	Electrospray ionization is a technique used in mass spectrometry to produce ions. It is especially useful in producing ions from macromolecules because it overcomes the propensity of these molecules to fragment when ionized. The development of electrospray ionization for the analysis of biological macromolecules was rewarded with the attribution of the Nobel Prize in Chemistry to John Bennett Fenn in 2002. One of the original instruments used by Dr. Fenn is on display at the Chemical Heritage Foundation in Philadelphia, Pennsylvania.
Fast atom bombardment	Fast atom bombardment is an ionization technique used in mass spectrometry. The material to be analyzed is mixed with a non-volatile chemical protection environment called a matrix and is bombarded under vacuum with a high energy (4000 to 10,000 electron volts) beam of atoms. The atoms are typically from an inert gas such as argon or xenon. Common matrices include glycerol, thioglycerol, 3-nitrobenzyl alcohol (3-NBA), 18-Crown-6 ether, 2-nitrophenyloctyl ether, sulfolane, diethanolamine, and triethanolamine. This technique is similar to secondary ion mass spectrometry and plasma desorption mass spectrometry.
Mass spectrum	A mass spectrum is an intensity vs. m/z (mass-to-charge ratio) plot representing a chemical analysis. Hence, the mass spectrum of a sample is a pattern representing the distribution of ions by mass (more correctly: mass-to-charge ratio) in a sample. It is a histogram usually acquired using an instrument called a mass spectrometer.
Molecular structure	The molecular structure of a substance is described by the combination of nuclei and electrons that comprise its constitute molecules. This includes the molecular geometry (essentially the arrangement, in space, of the equilibrium positions of the constituent atoms -- in reality, these are in a state of constant vibration, at temperatures above absolute zero), the electronic properties of the bonds, and further molecular properties. The determination of molecular structure uses a multitude of experimental methods, that include X-ray diffraction, electron diffraction, many kinds of optical spectroscopy, nuclear magnetic resonance, electron spin resonance, and mass spectrometry.
Coupling	A coupling is a device used to connect two shafts together at their ends for the purpose of transmitting power. Couplings do not normally allow disconnection of shafts during operation, however there are torque limiting couplings which can slip or disconnect when some torque limit is exceeded.

Chapter 11. Chemical analysis 1: how do we know what is there?

	The primary purpose of couplings is to join two pieces of rotating equipment while permitting some degree of misalignment or end movement or both.
Gas chromatography-mass spectrometry	Gas chromatography-mass spectrometry is a method that combines the features of gas-liquid chromatography and mass spectrometry to identify different substances within a test sample. Applications of GC-MS include drug detection, fire investigation, environmental analysis, explosives investigation, and identification of unknown samples. GC/MS can also be used in airport security to detect substances in luggage or on human beings.
Liquid chromatography-mass spectrometry	Liquid chromatography-mass spectrometry is an analytical chemistry technique that combines the physical separation capabilities of liquid chromatography (or HPLC) with the mass analysis capabilities of mass spectrometry. LC-MS is a powerful technique used for many applications which has very high sensitivity and selectivity. Generally its application is oriented towards the specific detection and potential identification of chemicals in the presence of other chemicals (in a complex mixture).
Tandem mass spectrometry	Tandem mass spectrometry, involves multiple steps of mass spectrometry selection, with some form of fragmentation occurring in between the stages. Tandem MS instruments Multiple stages of mass analysis separation can be accomplished with individual mass spectrometer elements separated in space or using a single mass spectrometer with the MS steps separated in time. Tandem in space In tandem mass spectrometry in space, the separation elements are physically separated and distinct, although there is a physical connection between the elements to maintain high vacuum.
Spectroscopy	Spectroscopy is the study of the interaction between matter and radiated energy. Historically, spectroscopy originated through the study of visible light dispersed according to its wavelength, e.g., by a prism. Later the concept was expanded greatly to comprise any interaction with radiative energy as a function of its wavelength or frequency.

Chapter 11. Chemical analysis 1: how do we know what is there?

Electromagnetic radiation	Electromagnetic radiation is a form of energy emitted and absorbed by charged particles, which exhibits wave-like behavior as it travels through space. EMR has both electric and magnetic field components, which stand in a fixed ratio of intensity to each other, and which oscillate in phase perpendicular to each other and perpendicular to the direction of energy and wave propagation. In vacuum, electromagnetic radiation propagates at a characteristic speed, the speed of light.
Electromagnetic spectrum	The electromagnetic spectrum is the range of all possible frequencies of electromagnetic radiation. The 'electromagnetic spectrum' of an object is the characteristic distribution of electromagnetic radiation emitted or absorbed by that particular object. The electromagnetic spectrum extends from low frequencies used for modern radio communication to gamma radiation at the short-wavelength (high-frequency) end, thereby covering wavelengths from thousands of kilometers down to a fraction of the size of an atom.
Rotation	A rotation is a circular movement of an object around a center of rotation. A three-dimensional object rotates always around an imaginary line called a rotation axis. If the axis is within the body, and passes through its center of mass the body is said to rotate upon itself, or spin.
Vibration	Vibration is a mechanical phenomenon whereby oscillations occur about an equilibrium point. The oscillations may be periodic such as the motion of a pendulum or random such as the movement of a tire on a gravel road. Vibration is occasionally 'desirable'.
Transmission	A machine consists of a power source and a power transmission system, which provides controlled application of the power. Merriam-Webster defines transmission as: an assembly of parts including the speed-changing gears and the propeller shaft by which the power is transmitted from an engine to a live axle. Often transmission refers simply to the gearbox that uses gears and gear trains to provide speed and torque conversions from a rotating power source to another device.

Chapter 11. Chemical analysis 1: how do we know what is there?

Transmittance	In optics and spectroscopy, transmittance is the fraction of incident light at a specified wavelength that passes through a sample. A related term is absorptance, or absorption factor, which is the fraction of radiation absorbed by a sample at a specified wavelength. Occasionally one also hears the terms visible transmittance and visible absorptance (VA), which are the respective fractions for the spectrum of light visible radiation.
Magnetic field	A magnetic field is a mathematical description of the magnetic influence of electric currents and magnetic materials. The magnetic field at any given point is specified by both a direction and a magnitude ; as such it is a vector field. The magnetic field is most commonly defined in terms of the Lorentz force it exerts on moving electric charges.
Nuclear magnetic resonance	Nuclear magnetic resonance is a physical phenomenon in which magnetic nuclei in a magnetic field absorb and re-emit electromagnetic radiation. This energy is at a specific resonance frequency which depends on the strength of the magnetic field and the magnetic properties of the isotope of the atoms; in practical applications, the frequency is similar to VHF and UHF television broadcasts (60-1000 MHz). NMR allows the observation of specific quantum mechanical magnetic properties of the atomic nucleus.
Nuclear magnetic resonance spectroscopy	Nuclear magnetic resonance spectroscopy, most commonly known as NMR spectroscopy, is a research technique that exploits the magnetic properties of certain atomic nuclei to determine physical and chemical properties of atoms or the molecules in which they are contained. It relies on the phenomenon of nuclear magnetic resonance and can provide detailed information about the structure, dynamics, reaction state, and chemical environment of molecules. Most frequently, NMR spectroscopy is used by chemists and biochemists to investigate the properties of organic molecules, though it is applicable to any kind of sample that contains nuclei possessing spin.
Chemical shift	In nuclear magnetic resonance (NMR) spectroscopy, the chemical shift is the resonant frequency of a nucleus relative to a standard. Often the position and number of chemical shifts are diagnostic of the structure of a molecule. Chemical shifts are also used to describe signals in other forms of spectroscopy such as photoemission spectroscopy.

Chapter 11. Chemical analysis 1: how do we know what is there?

Functional magnetic resonance imaging	Functional magnetic resonance imaging is an MRI procedure that measures brain activity by detecting associated changes in blood flow. The primary form of fMRI uses the blood-oxygen-level-dependent (BOLD) contrast, discovered by Seiji Ogawa. This is a type of specialized brain and body scan used to map neural activity in the brain or spinal cord of humans or animals by imaging the change in blood flow (hemodynamic response) related to energy use by brain cells.
Magnetic resonance imaging	Magnetic resonance imaging nuclear magnetic resonance imaging or magnetic resonance tomography (MRT) is a medical imaging technique used in radiology to visualize internal structures of the body in detail. MRI makes use of the property of nuclear magnetic resonance (NMR) to image nuclei of atoms inside the body. An MRI scanner is a device in which the patient lies within a large, powerful magnet where the magnetic field is used to align the magnetization of some atomic nuclei in the body, and radio frequency fields to systematically alter the alignment of this magnetization.
Frequency	Frequency is the number of occurrences of a repeating event per unit time. It is also referred to as temporal frequency. The period is the duration of one cycle in a repeating event, so the period is the reciprocal of the frequency.
Infrared spectroscopy	Infrared spectroscopy is the spectroscopy that deals with the infrared region of the electromagnetic spectrum, that is light with a longer wavelength and lower frequency than visible light. It covers a range of techniques, mostly based on absorption spectroscopy. As with all spectroscopic techniques, it can be used to identify and study chemicals.
Wavenumber	In the physical sciences, the wavenumber is a property of a wave, its spatial frequency, that is proportional to the reciprocal of the wavelength. It is also the magnitude of the wave vector. Its usual symbols are ν, $\bar{\nu}$, σ or k, the first three used for one definition, the last for another.

Chapter 11. Chemical analysis 1: how do we know what is there?

| Diffraction | Diffraction refers to various phenomena which occur when a wave encounters an obstacle. Italian scientist Francesco Maria Grimaldi coined the word 'diffraction' and was the first to record accurate observations of the phenomenon in 1665. In classical physics, the diffraction phenomenon is described as the apparent bending of waves around small obstacles and the spreading out of waves past small openings. Similar effects occur when light waves travel through a medium with a varying refractive index or a sound wave through one with varying acoustic impedance. |

PRACTICE QUIZ
Chapter 11. Chemical analysis 1: how do we know what is there?

1. Agar or agar-agar is a gelatinous substance derived by boiling from a polysaccharide in red algae, where it accumulates in the cell walls of agarophyte and serve as the primary structural support for the algae's cell walls. Agar is a mixture of two components: the linear polysaccharide _____, and a heterogeneous mixture of smaller molecules called agaropectin.

    Throughout history into modern times, agar has been chiefly used as an ingredient in desserts throughout Asia and also as a solid substrate to contain culture medium for microbiological work.

    a. Alginic acid
    b. Alguronic acid
    c. Agarose
    d. Amylopectin

2. The _____ concentration or blood glucose level is the amount of glucose (sugar) present in the blood of a human or animal. Normally in mammals, the body maintains the blood glucose level at a reference range between about 3.6 and 5.8 mM (mmol/L, i.e., millimoles/liter), or 64.8 and 104.4 mg/dL. The human body naturally tightly regulates blood glucose levels as a part of metabolic homeostasis.

    Glucose is the primary source of energy for the body's cells, and blood lipids (in the form of fats and oils) are primarily a compact energy store.

    a. Blood test
    b. Blood sugar
    c. Bodansky unit
    d. C-reactive protein

3. _____ also known as electrofocusing, is a technique for separating different molecules by their electric charge differences. It is a type of zone electrophoresis, usually performed on proteins in a gel, that takes advantage of the fact that overall charge on the molecule of interest is a function of the pH of its surroundings.

    _____ in laboratory
    IEF involves adding an ampholyte solution into immobilized pH gradient (IPG) gels.

---

a. Aalto Vase
   b. Isopeptide bond
   c. Octet rule
   d. Isoelectric focusing

4. _____(taxonomy) - mainly in life sciences or natural history - refers to resemblance suggesting a common descent, phylogenetic relationship, or type. The term does, however, have broader application, such as in geology (for example, in descriptive and theoretical works), and similarly in astronomy .

   Basis
   In taxonomy the basis of any particular type of classification is the way in which objects in the domain resemble each other.

   a. Affinity
   b. Afossochitonidae
   c. Allochiton
   d. Alpha taxonomy

5. _____ is an ionization technique used in mass spectrometry. The material to be analyzed is mixed with a non-volatile chemical protection environment called a matrix and is bombarded under vacuum with a high energy (4000 to 10,000 electron volts) beam of atoms. The atoms are typically from an inert gas such as argon or xenon. Common matrices include glycerol, thioglycerol, 3-nitrobenzyl alcohol (3-NBA), 18-Crown-6 ether, 2-nitrophenyloctyl ether, sulfolane, diethanolamine, and triethanolamine. This technique is similar to secondary ion mass spectrometry and plasma desorption mass spectrometry.

   a. 1,2-Dioxetanedione
   b. Inductively coupled plasma
   c. Fast atom bombardment
   d. Volume fraction

ANSWER KEY
Chapter 11. Chemical analysis 1: how do we know what is there?

1. c
2. b
3. d
4. a
5. c

You can take the complete Chapter Practice Test

for Chapter 11. Chemical analysis 1: how do we know what is there?
on all key terms, persons, places, and concepts.

Online 99 Cents

http://www.epub14.51.19910.11.cram101.com/

Use www.Cram101.com for all your study needs

including Cram101's online interactive problem solving labs in chemistry, statistics, mathematics, and more.

CHAPTER OUTLINE: KEY TERMS, PEOPLE, PLACES, CONCEPTS
Chapter 12
Chemical analysis 2: how do we know how much is there?

- Mole
- Avogadro constant
- Molar mass
- Atomic mass
- Atomic mass unit
- DNA profiling
- Concentration
- Racemic mixture
- Solubility
- Solvent
- Gel electrophoresis
- Serial dilution
- Blood sugar
- Absorbance
- Spectrophotometer
- Spectrophotometry
- Reflection
- Molar absorptivity
- Path length

Chapter 12. Chemical analysis 2: how do we know how much is there?

- Atomic spectroscopy
- Energy transfer
- Vitamin A
- Absorption spectroscopy
- Atomic absorption spectroscopy
- Atomic emission spectroscopy
- Calibration curve
- Hollow cathode lamp
- Fluorescence
- Fluorescence spectroscopy
- Fluorophore
- High-performance liquid chromatography
- Green fluorescent protein
- Atomic weight
- Reporter gene
- Titration
- Coupling
- Environmental monitoring
- Phenolphthalein

Chapter 12. Chemical analysis 2: how do we know how much is there?

_____ | Silver chromate

_____ | Biosensor

_____ | Electrode

CHAPTER HIGHLIGHTS: KEY TERMS, PEOPLE, PLACES, CONCEPTS
Chapter 12. Chemical analysis 2: how do we know how much is there?

Mole	The mole is a unit of measurement for the amount of substance or chemical amount. It is one of the base units in the International System of Units, and has the unit symbol mol.
	The name mole is an 1897 translation of the German unit Mol, coined by the chemist Wilhelm Ostwald in 1893, although the related concept of equivalent mass had been in use at least a century earlier.
Avogadro constant	In chemistry and physics, the Avogadro constant is defined as the ratio of the number of constituent particles N (usually atoms or molecules) in a sample to the amount of substance n (unit mole) through the relationship $N_A = Nn$. Thus, it is the proportionality factor that relates the molar mass of an entity, i.e., the mass per amount of substance, to the mass of said entity. The Avogadro constant expresses the number of elementary entities per mole of substance and it has the value $6.022\ 141\ 29(27) \times 10^{23}$ mol^{-1}.
Molar mass	Molar mass, symbol M, is a physical property of a given substance (chemical element or chemical compound), namely its mass per amount of substance. The base SI unit for mass is the kilogram and that for amount of substance is the mole. Thus, the derived unit for molar mass is kg/mol.
Atomic mass	The atomic mass is the mass of a specific isotope, most often expressed in unified atomic mass units. The atomic mass is the total mass of protons, neutrons and electrons in a single atom.
	The atomic mass is sometimes incorrectly used as a synonym of relative atomic mass, average atomic mass and atomic weight; these differ subtly from the atomic mass.
Atomic mass unit	The unified atomic mass unit or dalton (symbol: Da) is a unit that is used for indicating mass on an atomic or molecular scale. It is defined as one twelfth of the rest mass of an unbound neutral atom of carbon-12 in its nuclear and electronic ground state, and has a value of $1.660\ 538\ 921(73) \times 10^{-27}$ kg. One dalton is approximately equal to the mass of one proton or one neutron; an equivalence of saying 1 g mol^{-1}.

Chapter 12. Chemical analysis 2: how do we know how much is there?

DNA profiling	DNA profiling is a technique employed by forensic scientists to assist in the identification of individuals by their respective DNA profiles. DNA profiles are encrypted sets of numbers that reflect a person's DNA makeup, which can also be used as the person's identifier. DNA profiling should not be confused with full genome sequencing.
Concentration	In chemistry, concentration is defined as the abundance of a constituent divided by the total volume of a mixture. Furthermore, in chemistry, four types of mathematical description can be distinguished: mass concentration, molar concentration, number concentration, and volume concentration. The term concentration can be applied to any kind of chemical mixture, but most frequently it refers to solutes in solutions.
Racemic mixture	In chemistry, a racemic mixture, is one that has equal amounts of left- and right-handed enantiomers of a chiral molecule. The first known racemic mixture was 'racemic acid', which Louis Pasteur found to be a mixture of the two enantiomeric isomers of tartaric acid. Nomenclature A racemic mixture is denoted by the prefix (±)- or dl- (for sugars the prefix DL- may be used), indicating an equal (1:1) mixture of dextro and levo isomers.
Solubility	Solubility is the property of a solid, liquid, or gaseous chemical substance called solute to dissolve in a solid, liquid, or gaseous solvent to form a homogeneous solution of the solute in the solvent. The solubility of a substance fundamentally depends on the used solvent as well as on temperature and pressure. The extent of the solubility of a substance in a specific solvent is measured as the saturation concentration, where adding more solute does not increase the concentration of the solution.
Solvent	A solvent is a liquid, solid, or gas that dissolves another solid, liquid, or gaseous solute, resulting in a solution that is soluble in a certain volume of solvent at a specified temperature. Common uses for organic solvents are in dry cleaning (e.g., tetrachloroethylene), as paint thinners (e.g., toluene, turpentine), as nail polish removers and glue solvents (acetone, methyl acetate, ethyl acetate), in spot removers (e.g., hexane, petrol ether), in detergents (citrus terpenes), in perfumes (ethanol), nail polish and in chemical synthesis. The use of inorganic solvents (other than water) is typically limited to research chemistry and some technological processes.

Chapter 12. Chemical analysis 2: how do we know how much is there?

Gel electrophoresis	Gel electrophoresis is a method used in clinical chemistry to separate proteins by charge and or size (IEF agarose, essentially size independent) and in biochemistry and molecular biology to separate a mixed population of DNA and RNA fragments by length, to estimate the size of DNA and RNA fragments or to separate proteins by charge. Nucleic acid molecules are separated by applying an electric field to move the negatively charged molecules through an agarose matrix. Shorter molecules move faster and migrate farther than longer ones because shorter molecules migrate more easily through the pores of the gel.
Serial dilution	A serial dilution is the stepwise dilution of a substance in solution. Usually the dilution factor at each step is constant, resulting in a geometric progression of the concentration in a logarithmic fashion. A ten-fold serial dilution could be 1 M, 0.1 M, 0.01 M, 0.001 M... Serial dilutions are used to accurately create highly diluted solutions as well as solutions for experiments resulting in concentration curves with a logarithmic scale.
Blood sugar	The blood sugar concentration or blood glucose level is the amount of glucose (sugar) present in the blood of a human or animal. Normally in mammals, the body maintains the blood glucose level at a reference range between about 3.6 and 5.8 mM (mmol/L, i.e., millimoles/liter), or 64.8 and 104.4 mg/dL. The human body naturally tightly regulates blood glucose levels as a part of metabolic homeostasis. Glucose is the primary source of energy for the body's cells, and blood lipids (in the form of fats and oils) are primarily a compact energy store.
Absorbance	In spectroscopy, the absorbance of a material is a logarithmic ratio of the radiation falling upon a material, to the radiation transmitted through a material. Absorbance measurements are often carried out in analytical chemistry. In physics, the term spectral absorbance is used interchangeably with spectral absorptance or absorptivity.

Chapter 12. Chemical analysis 2: how do we know how much is there?

Spectrophotometer	In chemistry, spectrophotometry is the quantitative measurement of the reflection or transmission properties of a material as a function of wavelength. It is more specific than the general term electromagnetic spectroscopy in that spectrophotometry deals with visible light, near-ultraviolet, and near-infrared, but does not cover time-resolved spectroscopic techniques. Spectrophotometry involves the use of a spectrophotometer.
Reflection	Reflection is the change in direction of a wavefront at an interface between two different media so that the wavefront returns into the medium from which it originated. Common examples include the reflection of light, sound and water waves. The law of reflection says that for specular reflection the angle at which the wave is incident on the surface equals the angle at which it is reflected.
Molar absorptivity	The molar absorption coefficient, molar extinction coefficient, or molar absorptivity, is a measurement of how strongly a chemical species absorbs light at a given wavelength. It is an intrinsic property of the species; the actual absorbance, A, of a sample is dependent on the pathlength, l, and the concentration, c, of the species via the Beer-Lambert law, $A = \epsilon c l$. The SI units for ϵ are m^2/mol, but in practice, they are usually taken as $M^{-1} cm^{-1}$ or $L\, mol^{-1} cm^{-1}$.
Path length	In chemistry, the path length is defined as the distance that light (UV/VIS) travels through a sample in an analytical cell. Typically, a sample cell is made of quartz, glass, or a plastic rhombic cuvette with a volume typically ranging from 0.1 mL to 10 mL or larger used in a spectrophotometer. For the purposes of spectrophotometry (i.e. when making calculations using the Beer-Lambert law) the path length is measured in centimeters (rather than in meters).
Atomic spectroscopy	Atomic spectroscopy is the determination of elemental composition by its electromagnetic or mass spectrum. Atomic spectroscopy is closely related to other forms of spectroscopy. It can be divided by atomization source or by the type of spectroscopy used.

Chapter 12. Chemical analysis 2: how do we know how much is there?

Energy transfer	Energy transfer is the transfer of energy from one object or material to another. There are a few main ways that energy transfer occurs: • Radiant energy (radiation) • Heat conduction (travelling heat) • Convection (currents of warm air) • Electrical power transmission • Mechanical work (machines)
Vitamin A	Vitamin A is a vitamin that is needed by the retina of the eye in the form of a specific metabolite, the light-absorbing molecule retinal, that is absolutely necessary for both low-light (scotopic vision) and color vision. Vitamin A also functions in a very different role, as an irreversibly oxidized form of retinol known as retinoic acid, which is an important hormone-like growth factor for epithelial and other cells. In foods of animal origin, the major form of vitamin A is an ester, primarily retinyl palmitate, which is converted to the retinol (chemically an alcohol) in the small intestine.
Absorption spectroscopy	Absorption spectroscopy refers to spectroscopic techniques that measure the absorption of radiation, as a function of frequency or wavelength, due to its interaction with a sample. The sample absorbs energy, i.e., photons, from the radiating field. The intensity of the absorption varies as a function of frequency, and this variation is the absorption spectrum.
Atomic absorption spectroscopy	Atomic absorption spectroscopy is a spectroanalytical procedure for the quantitative determination of chemical elements employing the absorption of optical radiation (light) by free atoms in the gaseous state. In analytical chemistry the technique is used for determining the concentration of a particular element (the analyte) in a sample to be analyzed. AAS can be used to determine over 70 different elements in solution or directly in solid samples.

Chapter 12. Chemical analysis 2: how do we know how much is there?

Atomic emission spectroscopy	Atomic emission spectroscopy is a method of chemical analysis that uses the intensity of light emitted from a flame, plasma, arc, or spark at a particular wavelength to determine the quantity of an element in a sample. The wavelength of the atomic spectral line gives the identity of the element while the intensity of the emitted light is proportional to the number of atoms of the element. Flame emission spectroscopy A sample of a material (analyte) is brought into the flame as a gas or sprayed solution.
Calibration curve	In analytical chemistry, a calibration curve is a general method for determining the concentration of a substance in an unknown sample by comparing the unknown to a set of standard samples of known concentration. A calibration curve is one approach to the problem of instrument calibration; other approaches may mix the standard into the unknown, giving an internal standard. The calibration curve is a plot of how the instrumental response, the so-called analytical signal, changes with the concentration of the analyte (the substance to be measured).
Hollow cathode lamp	A hollow cathode lamp is type of lamp used in physics and chemistry as a spectral line source (e.g. for atomic absorption spectrometers) and as a frequency tuner for light sources such as lasers. An Hollow cathode lamp usually consists of a glass tube containing a cathode, an anode, and a buffer gas (usually a noble gas). A large voltage across the anode and cathode will cause the buffer gas to ionize, creating a plasma.
Fluorescence	Fluorescence is the emission of light by a substance that has absorbed light or other electromagnetic radiation. It is a form of luminescence. In most cases, the emitted light has a longer wavelength, and therefore lower energy, than the absorbed radiation.

Chapter 12. Chemical analysis 2: how do we know how much is there?

Fluorescence spectroscopy	Fluorescence spectroscopy aka fluorometry or spectrofluorometry, is a type of electromagnetic spectroscopy which analyzes fluorescence from a sample. It involves using a beam of light, usually ultraviolet light, that excites the electrons in molecules of certain compounds and causes them to emit light; typically, but not necessarily, visible light. A complementary technique is absorption spectroscopy.
Fluorophore	A fluorophore is a fluorescent chemical compound that can re-emit light upon light excitation. Fluorophores typically contain several combined aromatic groups, or plane or cyclic molecules with several π bonds.
	Fluorophores are sometimes used alone, as a tracer in fluids, as a dye for staining of certain structures, as a substrate of enzymes, or as a probe or indicator (when its fluorescence is affected by environment such as polarity, ions,.)...
High-performance liquid chromatography	High-performance liquid chromatography HPLC, is a chromatographic technique used to separate a mixture of compounds in analytical chemistry and biochemistry with the purpose of identifying, quantifying and purifying the individual components of the mixture. Some common examples are the separation and quantitation of performance enhancement drugs (e.g. steroids) in urine samples, or of vitamin D levels in serum.
	HPLC typically utilizes different types of stationary phases (i.e. sorbents) contained in columns, a pump that moves the mobile phase and sample components through the column, and a detector capable of providing characteristic retention times for the sample components and area counts reflecting the amount of each analyte passing through the detector.
Green fluorescent protein	The green fluorescent protein is a protein composed of 238 amino acid residues (26.9kDa) that exhibits bright green fluorescence when exposed to light in the blue to ultraviolet range. Although many other marine organisms have similar green fluorescent proteins, GFP traditionally refers to the protein first isolated from the jellyfish Aequorea victoria. The GFP from A. victoria has a major excitation peak at a wavelength of 395 nm and a minor one at 475 nm.

Chapter 12. Chemical analysis 2: how do we know how much is there?

Atomic weight	Atomic weight is a dimensionless physical quantity, the ratio of the average mass of atoms of an element (from a given source) to 1/12 of the mass of an atom of carbon-12 (known as the unified atomic mass unit). The term is usually used, without further qualification, to refer to the standard atomic weights published at regular intervals by the International Union of Pure and Applied Chemistry (IUPAC) and which are intended to be applicable to normal laboratory materials.
Reporter gene	In molecular biology, a reporter gene is a gene that researchers attach to a regulatory sequence of another gene of interest in bacteria, cell culture, animals or plants. Certain genes are chosen as reporters because the characteristics they confer on oanisms expressing them are easily identified and measured, or because they are selectable markers. Reporter genes are often used as an indication of whether a certain gene has been taken up by or expressed in the cell or oanism population.
Titration	Titration, is a common laboratory method of quantitative chemical analysis that is used to determine the unknown concentration of an identified analyte. Because volume measurements play a key role in titration, it is also known as volumetric analysis. A reagent, called the titrant or titrator is prepared as a standard solution.
Coupling	A coupling is a device used to connect two shafts together at their ends for the purpose of transmitting power. Couplings do not normally allow disconnection of shafts during operation, however there are torque limiting couplings which can slip or disconnect when some torque limit is exceeded. The primary purpose of couplings is to join two pieces of rotating equipment while permitting some degree of misalignment or end movement or both.
Environmental monitoring	Environmental monitoring describes the processes and activities that need to take place to characterise and monitor the quality of the environment. Environmental monitoring is used in the preparation of environmental impact assessments, as well as in many circumstances in which human activities carry a risk of harmful effects on the natural environment. All monitoring strategies and programmes have reasons and justifications which are often designed to establish the current status of an environment or to establish trends in environmental parameters.

Chapter 12. Chemical analysis 2: how do we know how much is there?

Phenolphthalein	Phenolphthalein is a chemical compound with the formula $C_{20}H_{14}O_4$ and is often written as 'HIn' or 'phph' in shorthand notation. Often used in titrations, it turns colorless in acidic solutions and pink in basic solutions. If the concentration of indicator is particularly strong, it can appear purple.
Silver chromate	Silver chromate is a brown-red monoclinic crystal and is a chemical precursor to modern photography. It can be formed by combining silver nitrate ($AgNO_3$) and potassium chromate (K_2CrO_4) or sodium chromate (Na_2CrO_4). This reaction has been important in neuroscience, as it is used in the 'Golgi method' of staining neurons for microscopy: the silver chromate produced precipitates inside neurons and makes their morphology visible.
Biosensor	A biosensor is an analytical device for the detection of an analyte that combines a biological component with a physicochemical detector component. It consists of 3 parts: - the sensitive biological element (biological material (e.g. tissue, microorganisms, organelles, cell receptors, enzymes, antibodies, nucleic acids, etc)., a biologically derived material or biomimic) The sensitive elements can be created by biological engineering. - the transducer or the detector element (works in a physicochemical way; optical, piezoelectric, electrochemical, etc). that transforms the signal resulting from the interaction of the analyte with the biological element into another signal (i.e., transducers) that can be more easily measured and quantified; - associated electronics or signal processors that are primarily responsible for the display of the results in a user-friendly way.
Electrode	An electrode is an electrical conductor used to make contact with a nonmetallic part of a circuit (e.g. a semiconductor, an electrolyte or a vacuum). The word was coined by the scientist Michael Faraday from the Greek words elektron (meaning amber, from which the word electricity is derived) and hodos, a way. Anode and cathode in electrochemical cells An electrode in an electrochemical cell is referred to as either an anode or a cathode (words that were also coined by Faraday).

PRACTICE QUIZ
Chapter 12. Chemical analysis 2: how do we know how much is there?

1. A _____ is a device used to connect two shafts together at their ends for the purpose of transmitting power. _____s do not normally allow disconnection of shafts during operation, however there are torque limiting _____s which can slip or disconnect when some torque limit is exceeded.

   The primary purpose of _____s is to join two pieces of rotating equipment while permitting some degree of misalignment or end movement or both.

   a. Crank
   b. Critical speed
   c. Cryogenic engineering
   d. Coupling

2. _____, symbol M, is a physical property of a given substance (chemical element or chemical compound), namely its mass per amount of substance. The base SI unit for mass is the kilogram and that for amount of substance is the mole. Thus, the derived unit for _____ is kg/mol.

   a. 1,2-Dioxetanedione
   b. Osmolarity
   c. Molar mass
   d. Osmotic pressure

3. In chemistry, _____ is the quantitative measurement of the reflection or transmission properties of a material as a function of wavelength. It is more specific than the general term electromagnetic spectroscopy in that _____ deals with visible light, near-ultraviolet, and near-infrared, but does not cover time-resolved spectroscopic techniques.

   _____ involves the use of a spectrophotometer.

   a. Spectroscopic notation
   b. Spectrophotometry
   c. Spin label
   d. Spin polarization

---

4. _____ is a method used in clinical chemistry to separate proteins by charge and or size (IEF agarose, essentially size independent) and in biochemistry and molecular biology to separate a mixed population of DNA and RNA fragments by length, to estimate the size of DNA and RNA fragments or to separate proteins by charge. Nucleic acid molecules are separated by applying an electric field to move the negatively charged molecules through an agarose matrix. Shorter molecules move faster and migrate farther than longer ones because shorter molecules migrate more easily through the pores of the gel.

   a. Gel electrophoresis
   b. Liquid gas
   c. Microscale thermophoresis
   d. Relative fluorescence units

5. A _____ is an analytical device for the detection of an analyte that combines a biological component with a physicochemical detector component.

   It consists of 3 parts:

   - the sensitive biological element (biological material (e.g. tissue, microorganisms, organelles, cell receptors, enzymes, antibodies, nucleic acids, etc)., a biologically derived material or biomimic) The sensitive elements can be created by biological engineering.
   - the transducer or the detector element (works in a physicochemical way; optical, piezoelectric, electrochemical, etc). that transforms the signal resulting from the interaction of the analyte with the biological element into another signal (i.e., transducers) that can be more easily measured and quantified;
   - associated electronics or signal processors that are primarily responsible for the display of the results in a user-friendly way.

   a. Biosensor
   b. Silver iodide
   c. Silver nitrate
   d. Sodium acetate

ANSWER KEY
Chapter 12. Chemical analysis 2: how do we know how much is there?

1. d
2. c
3. b
4. a
5. a

You can take the complete Chapter Practice Test

for Chapter 12. Chemical analysis 2: how do we know how much is there?
on all key terms, persons, places, and concepts.

Online 99 Cents

http://www.epub14.51.19910.12.cram101.com/

Use www.Cram101.com for all your study needs

including Cram101's online interactive problem solving labs in chemistry, statistics, mathematics, and more.

CHAPTER OUTLINE: KEY TERMS, PEOPLE, PLACES, CONCEPTS
Chapter 13
Energy: what makes reactions go?

_____ DNA profiling

_____ Electron pair

_____ Internal energy

_____ Kinetic energy

_____ Potential

_____ Potential energy

_____ Velocity

_____ Chemical element

_____ Chemical energy

_____ Gravitational energy

_____ Bond energy

_____ Covalent bond

_____ Ionic bond

_____ Bond cleavage

_____ Closed system

_____ Energy transfer

_____ Intramolecular force

_____ Isolated system

_____ Open system

Chapter 13. Energy: what makes reactions go?

- Thermal energy
- Gravitational field
- Thermal
- Chemical reaction
- Exothermic
- Exothermic reaction
- Endothermic
- Racemic mixture
- Standard state
- Calorimetry
- Combustion
- Metabolism
- Universe
- Refrigeration
- Gibbs free energy
- Endergonic reaction
- Exergonic
- Exergonic reaction
- Respiration

Chapter 13. Energy: what makes reactions go?

Hydrolysis

CHAPTER HIGHLIGHTS: KEY TERMS, PEOPLE, PLACES, CONCEPTS
Chapter 13. Energy: what makes reactions go?

DNA profiling	DNA profiling is a technique employed by forensic scientists to assist in the identification of individuals by their respective DNA profiles. DNA profiles are encrypted sets of numbers that reflect a person's DNA makeup, which can also be used as the person's identifier. DNA profiling should not be confused with full genome sequencing.
Electron pair	In chemistry, an electron pair consists of two electrons that occupy the same orbital but have opposite spins.
	Because electrons are fermions, the Pauli exclusion principle forbids these particles from having exactly the same quantum numbers. Therefore the only way to occupy the same orbital, i.e. have the same orbital quantum numbers, is to differ in the spin quantum number.
Internal energy	In thermodynamics, the internal energy is the total energy contained by a thermodynamic system. It is the energy needed to create the system, but excludes the energy to displace the system's surroundings, any energy associated with a move as a whole, or due to external force fields. Internal energy has two major components, kinetic energy and potential energy.
Kinetic energy	The kinetic energy of an object is the energy which it possesses due to its motion. It is defined as the work needed to accelerate a body of a given mass from rest to its stated velocity. Having gained this energy during its acceleration, the body maintains this kinetic energy unless its speed changes.
Potential	- In linguistics, the potential mood
- The mathematical study of potentials is known as potential theory; it is the study of harmonic functions on manifolds. This mathematical formulation arises from the fact that, in physics, the scalar potential is irrotational, and thus has a vanishing Laplacian -- the very definition of a harmonic function.
- In physics, a potential may refer to the scalar potential or to the vector potential. In either case, it is a field defined in space, from which many important physical properties may be derived.
Potential energy	In physics, potential energy is the energy of a body or a system due to the position of the body or the arrangement of the particles of the system. The SI unit for measuring work and energy is the Joule (symbol J).

Chapter 13. Energy: what makes reactions go?

	The term 'potential energy' was coined by the 19th century Scottish engineer and physicist William Rankine.
Velocity	In physics, velocity is speed in a given direction. Speed describes only how fast an object is moving, whereas velocity gives both the speed and direction of the object's motion. To have a constant velocity, an object must have a constant speed and motion in a constant direction.
Chemical element	A chemical element is a pure chemical substance consisting of one type of atom distinguished by its atomic number, which is the number of protons in its nucleus. They are divided into metals and non-metals. Familiar examples of elements include carbon, oxygen (non-metals) together with aluminium, iron, copper, gold, mercury, and lead (metals).
Chemical energy	In chemistry, Chemical energy is the potential of a chemical substance to undergo a transformation through a chemical reaction or, to transform other chemical substances. Breaking or making of chemical bonds involves energy, which may be either absorbed or evolved from a chemical system. Energy that can be released because of a reaction between a set of chemical substances is equal to the difference between the energy content of the products and the reactants.
Gravitational energy	Gravitational energy is the energy associated with the gravitational field. This phrase is found frequently in scientific writings about quasars (quasi-stellar objects) and other active galaxies. Newtonian mechanics According to classical mechanics, between two or more masses a gravitational potential energy exists, from which the gravitational field energy density can be calculated.
Bond energy	In chemistry, bond energy is the measure of bond strength in a chemical bond. It is the heat required to break one Mole (unit) of molecules into their individual atoms. For example, the carbon-hydrogen bond energy in methane E(C-H) is the enthalpy change involved with breaking up one molecule of methane into a carbon atom and 4 hydrogen radicals divided by 4. Bond energy should not be confused with bond-dissociation energy. Bond energy/distance correlation

---

Chapter 13. Energy: what makes reactions go?

	Bond strength (energy) can be directly related to the bond length/bond distance.
Covalent bond	A covalent bond is a form of chemical bonding that is characterized by the sharing of pairs of electrons between atoms. The stable balance of attractive and repulsive forces between atoms when they share electrons is known as covalent bonding.
	Covalent bonding includes many kinds of interaction, including σ-bonding, π-bonding, metal-to-metal bonding, agostic interactions, and three-center two-electron bonds.
Ionic bond	An ionic bond is a type of chemical bond formed through an electrostatic attraction between two oppositely charged ions. Ionic bonds are formed between a cation, which is usually a metal, and an anion, which is usually a nonmetal. Pure ionic bonding cannot exist: all ionic compounds have some degree of covalent bonding.
Bond cleavage	Bond cleavage, is the splitting of chemical bonds.
	If the two electrons in a cleaved covalent bond are divided between the products, the process is known as homolytic fission or homolysis and free redicals are generated by homolytic cleavage. Alternatively, the case where both electrons are retained by one product and charged species that is nucleophile and electrophile are generated by the process is known as heterolytic fission and (heterolysis).
Closed system	The term closed system has different meanings in different contexts. In thermodynamics In thermodynamics, a closed system can exchange energy (as heat or work), but not matter, with its surroundings. In contrast, an isolated system cannot exchange any of heat, work, or matter with the surroundings, while an open system can exchange all of heat, work and matter.

Chapter 13. Energy: what makes reactions go?

Energy transfer	Energy transfer is the transfer of energy from one object or material to another. There are a few main ways that energy transfer occurs: - Radiant energy (radiation) - Heat conduction (travelling heat) - Convection (currents of warm air) - Electrical power transmission - Mechanical work (machines)
Intramolecular force	An intramolecular force is any force that holds together the atoms making up a molecule or compound. They contain all types of chemical bond. They are stronger than intermolecular forces, which are present between atoms or molecules that are not actually bonded.
Isolated system	In the natural sciences an isolated system is a physical system without any external exchange - neither matter nor energy can enter or exit, but can only move around inside. Truly isolated systems cannot exist in nature, other than possibly the universe itself, and they are thus hypothetical concepts only. It obeys in particular the first of the conservation laws: its total energy - mass stays constant.
Open system	An open system is a system which continuously interacts with its environment. An open system should be contrasted with the concept of an isolated system which exchanges neither energy, matter, nor information with its environment. The concept of an 'open system' was formalized within a framework that enabled one to interrelate the theory of the organism, thermodynamics, and evolutionary theory.

Chapter 13. Energy: what makes reactions go?

Thermal energy	Thermal energy is the part of the total internal energy of a thermodynamic system or sample of matter that results in the system temperature. The internal energy, also often called the thermodynamic energy, includes other forms of energy in a thermodynamic system in addition to thermal energy, namely forms of potential energy that do not influence temperature, such as the chemical energy stored in its molecular structure and electronic configuration, intermolecular interactions associated with phase changes that do not influence temperature (i.e., latent energy), and the nuclear binding energy that binds the sub-atomic particles of matter. Microscopically, the thermal energy is partly the kinetic energy of a system's constituent particles, which may be atoms, molecules, electrons, or particles in plasmas.
Gravitational field	The gravitational field is a model used in physics to explain the existence of gravity. In its original concept, gravity was a force between point masses. Following Newton, Laplace attempted to model gravity as some kind of radiation field or fluid, and since the 19th century explanations for gravity have usually been sought in terms of a field model, rather than a point attraction.
Thermal	A thermal column is a column of rising air in the lower altitudes of the Earth's atmosphere. Thermals are created by the uneven heating of the Earth's surface from solar radiation, and are an example of convection, specifically atmospheric convection. The Sun warms the ground, which in turn warms the air directly above it.
Chemical reaction	A chemical reaction is a process that leads to the transformation of one set of chemical substances to another. Chemical reactions can be either spontaneous, requiring no input of energy, or non-spontaneous, typically following the input of some type of energy, such as heat, light or electricity. Classically, chemical reactions encompass changes that strictly involve the motion of electrons in the forming and breaking of chemical bonds, although the general concept of a chemical reaction, in particular the notion of a chemical equation, is applicable to transformations of elementary particles (such as illustrated by Feynman diagrams), as well as nuclear reactions.
Exothermic	In thermodynamics, the term exothermic describes a process or reaction that releases energy from the system, usually in the form of heat, but also in the form of light (e.g. a spark, flame, or explosion), electricity (e.g. a battery), or sound (e.g. burning hydrogen). Its etymology stems from the prefix exo and the Greek word thermasi (meaning 'to heat'). The term exothermic was first coined by Marcellin Berthelot.

Chapter 13. Energy: what makes reactions go?

Exothermic reaction	An exothermic reaction is a chemical reaction that releases energy in the form of light or heat. It is the opposite of an endothermic reaction. Expressed in a chemical equation: reactants → products + energy Overview An exothermic reaction is a chemical reaction that is accompanied by the release of heat.
Endothermic	In thermodynamics, the word endothermic describes a process or reaction in which the system absorbs energy from the surroundings in the form of heat. It is a modern coinage formed from Greek roots (as is often the case with scientific terminology). The prefix endo- derives from the Greek word 'endon' (?νδον) meaning 'within,' and the latter part of the word comes from the Greek word root 'therm' (θερμ-) meaning 'hot.' Hence it refers to a reaction that needs heat.
Racemic mixture	In chemistry, a racemic mixture, is one that has equal amounts of left- and right-handed enantiomers of a chiral molecule. The first known racemic mixture was 'racemic acid', which Louis Pasteur found to be a mixture of the two enantiomeric isomers of tartaric acid. Nomenclature A racemic mixture is denoted by the prefix (±)- or dl- (for sugars the prefix DL- may be used), indicating an equal (1:1) mixture of dextro and levo isomers.
Standard state	In chemistry, the standard state of a material (pure substance, mixture or solution) is a reference point used to calculate its properties under different conditions. In principle, the choice of standard state is arbitrary, although the International Union of Pure and Applied Chemistry (IUPAC) recommends a conventional set of standard states for general use. IUPAC recommends using a standard pressure $p^{\ominus}$ = 1 bar (100 kilopascals).
Calorimetry	Calorimetry is the science of measuring the heat of chemical reactions or physical changes. Calorimetry is performed with a calorimeter. The word calorimetry is derived from the Latin word calor, meaning heat.

Chapter 13. Energy: what makes reactions go?

Combustion	Combustion or burning is the sequence of exothermic chemical reactions between a fuel and an oxidant accompanied by the production of heat and conversion of chemical species. The release of heat can result in the production of light in the form of either glowing or a flame. Fuels of interest often include organic compounds (especially hydrocarbons) in the gas, liquid or solid phase.
Metabolism	Metabolism is the set of chemical reactions that happen in the cells of living organisms to sustain life. These processes allow organisms to grow and reproduce, maintain their structures, and respond to their environments. The word metabolism can also refer to all chemical reactions that occur in living organisms, including digestion and the transport of substances into and between different cells, in which case the set of reactions within the cells is called intermediary metabolism or intermediate metabolism.
Universe	The universe is commonly defined as the totality of everything that exists, including all matter and energy, the planets, stars, galaxies, and the contents of intergalactic space. Definitions and usage vary and similar terms include the cosmos, the world and nature. Scientific observation of earlier stages in the development of the universe, which can be seen at great distances, suggests that the universe has been governed by the same physical laws and constants throughout most of its extent and history.
Refrigeration	Refrigeration is a process in which work is done to move heat from one location to another. The work of heat transport is traditionally driven by mechanical work, but can also be driven by magnetism, laser or other means. Refrigeration has many applications, including, but not limited to: household refrigerators, industrial freezers, cryogenics, air conditioning, and heat pumps.
Gibbs free energy	Property database In thermodynamics, the Gibbs free energy is a thermodynamic potential that measures the 'useful' or process-initiating work obtainable from a thermodynamic system at a constant temperature and pressure (isothermal, isobaric). Just as in mechanics, where potential energy is defined as capacity to do work, similarly different potentials have different meanings. The Gibbs free energy is the maximum amount of non-expansion work that can be extracted from a closed system; this maximum can be attained only in a completely reversible process.

Chapter 13. Energy: what makes reactions go?

Endergonic reaction	In chemical thermodynamics, an endergonic reaction is a chemical reaction in which the standard change in free energy is positive, and energy is absorbed. In layman's terms the total amount of energy is a loss (it takes more energy to start the reaction than what you get out of it) so the total energy is a negative net result. Under constant temperature and constant pressure conditions, this means that the change in the standard Gibbs free energy would be positive $$\Delta G° > 0$$ for the reaction at standard state (i.e. at standard pressure (1 bar), and standard concentrations (1 molar) of all the reagents).
Exergonic	Exergonic means 'releasing energy in the form of work'. In thermodynamics, work is defined as the energy moving from the system (the internal region) to the surroundings (the external region) during a given process. An exergonic process is one in which there is a positive flow of energy from the system to the surroundings.
Exergonic reaction	An exergonic reaction is a chemical reaction where the change in the Gibbs free energy is negative, indicating a spontaneous reaction. Symbolically, the release of Gibbs free energy, G, in an exergonic reaction is denoted as $$\Delta G = G_{products} - G_{reactants} < 0.$$ Although exergonic reactions are said to occur spontaneously, this does not imply that the reaction will take place at an observable rate. For instance, the disproportionation of hydrogen peroxide is very slow in the absence of a suitable catalyst.

Chapter 13. Energy: what makes reactions go?

Respiration	In physiology, respiration (often confused with breathing) is defined as the transport of oxygen from the outside air to the cells within tissues, and the transport of carbon dioxide in the opposite direction. This is in contrast to the biochemical definition of respiration, which refers to cellular respiration: the metabolic process by which an organism obtains energy by reacting oxygen with glucose to give water, carbon dioxide and ATP (energy). Although physiologic respiration is necessary to sustain cellular respiration and thus life in animals, the processes are distinct: cellular respiration takes place in individual cells of the organism, while physiologic respiration concerns the bulk flow and transport of metabolites between the organism and the external environment.
Hydrolysis	Hydrolysis usually means the rupture of chemical bonds by the addition of water. Generally, hydrolysis is a step in the degradation of a substance. In terms of the word's derivation, hydrolysis comes from Greek roots hydro 'water' + lysis 'separation'.

PRACTICE QUIZ
Chapter 13. Energy: what makes reactions go?

1. In physics, _____ is speed in a given direction. Speed describes only how fast an object is moving, whereas _____ gives both the speed and direction of the object's motion. To have a constant _____, an object must have a constant speed and motion in a constant direction.

   a. 1,2-Dioxetanedione
   b. Velocity
   c. Proper velocity
   d. Relative density

2. A _____ is a pure chemical substance consisting of one type of atom distinguished by its atomic number, which is the number of protons in its nucleus. They are divided into metals and non-metals. Familiar examples of elements include carbon, oxygen (non-metals) together with aluminium, iron, copper, gold, mercury, and lead (metals).

   a. Chemical energy
   b. Chemical library
   c. Chemical element
   d. Chemical similarity

3. A _____ is a form of chemical bonding that is characterized by the sharing of pairs of electrons between atoms. The stable balance of attractive and repulsive forces between atoms when they share electrons is known as covalent bonding.

   Covalent bonding includes many kinds of interaction, including σ-bonding, π-bonding, metal-to-metal bonding, agostic interactions, and three-center two-electron bonds.

   a. Covalent Bond Classification
   b. Criegee biradical
   c. Cross-conjugation
   d. Covalent bond

4. In the natural sciences an _____ is a physical system without any external exchange - neither matter nor energy can enter or exit, but can only move around inside. Truly _____s cannot exist in nature, other than possibly the universe itself, and they are thus hypothetical concepts only. It obeys in particular the first of the conservation laws: its total energy - mass stays constant.

---

a. Isolated system
b. Isopeptide bond
c. Octet rule
d. Open shell

5. _____, is the splitting of chemical bonds.

If the two electrons in a cleaved covalent bond are divided between the products, the process is known as homolytic fission or homolysis and free redicals are generated by homolytic cleavage. Alternatively, the case where both electrons are retained by one product and charged species that is nucleophile and electrophile are generated by the process is known as heterolytic fission and (heterolysis).

a. Bond energy
b. Bond order
c. Bond valence method
d. Bond cleavage

ANSWER KEY
Chapter 13. Energy: what makes reactions go?

1. b
2. c
3. d
4. a
5. d

You can take the complete Chapter Practice Test

for Chapter 13. Energy: what makes reactions go?
on all key terms, persons, places, and concepts.

Online 99 Cents

http://www.epub14.51.19910.13.cram101.com/

Use www.Cram101.com for all your study needs

including Cram101's online interactive problem solving labs in chemistry, statistics, mathematics, and more.

CHAPTER OUTLINE: KEY TERMS, PEOPLE, PLACES, CONCEPTS
Chapter 14
Kinetics: what affects the speed of a reaction?

- _____ Racemic mixture
- _____ DNA-binding protein
- _____ Carbon monoxide
- _____ Disulfide bond
- _____ DNA profiling
- _____ Reaction rate
- _____ Warfarin
- _____ Order
- _____ Order of reaction
- _____ Collision theory
- _____ Avogadro constant
- _____ Pasteurization
- _____ Activation energy
- _____ Transition state
- _____ Catalysis
- _____ Catalytic converter
- _____ Enzyme
- _____ Ribozyme
- _____ Active site

Chapter 14. Kinetics: what affects the speed of a reaction?

_____ Catalase

_____ Enzyme catalysis

_____ Enzyme kinetics

_____ Hexokinase

_____ Phenylketonuria

_____ Electron pair

_____ Turnover number

CHAPTER HIGHLIGHTS: KEY TERMS, PEOPLE, PLACES, CONCEPTS
Chapter 14. Kinetics: what affects the speed of a reaction?

Racemic mixture	In chemistry, a racemic mixture, is one that has equal amounts of left- and right-handed enantiomers of a chiral molecule. The first known racemic mixture was 'racemic acid', which Louis Pasteur found to be a mixture of the two enantiomeric isomers of tartaric acid. Nomenclature A racemic mixture is denoted by the prefix (±)- or dl- (for sugars the prefix DL- may be used), indicating an equal (1:1) mixture of dextro and levo isomers.
DNA-binding protein	DNA-binding proteins are proteins that are composed of DNA-binding domains and thus have a specific or general affinity for either single or double stranded DNA. Sequence-specific DNA-binding proteins generally interact with the major groove of B-DNA, because it exposes more functional groups that identify a base pair. However there are some known minor groove DNA-binding ligands such as Netropsin, Distamycin, Hoechst 33258, Pentamidine and others. Examples DNA-binding proteins include transcription factors which modulate the process of transcription, various polymerases, nucleases which cleave DNA molecules, and histones which are involved in chromosome packaging and transcription in the cell nucleus.
Carbon monoxide	Carbon monoxide also called carbonous oxide, is a colorless, odorless, and tasteless gas which is slightly lighter than air. It is highly toxic to humans and animals in higher quantities, although it is also produced in normal animal metabolism in low quantities, and is thought to have some normal biological functions. Carbon monoxide consists of one carbon atom and one oxygen atom, connected by a triple bond which consists of two covalent bonds as well as one dative covalent bond.
Disulfide bond	In chemistry, a disulfide bond is a covalent bond, usually derived by the coupling of two thiol groups. The linkage is also called an SS-bond or disulfide bridge. The overall connectivity is therefore R-S-S-R. The terminology is widely used in biochemistry.

Chapter 14. Kinetics: what affects the speed of a reaction?

DNA profiling	DNA profiling is a technique employed by forensic scientists to assist in the identification of individuals by their respective DNA profiles. DNA profiles are encrypted sets of numbers that reflect a person's DNA makeup, which can also be used as the person's identifier. DNA profiling should not be confused with full genome sequencing.
Reaction rate	The reaction rate or speed of reaction for a reactant or product in a particular reaction is intuitively defined as how fast or slow a reaction takes place. For example, the oxidation of iron under the atmosphere is a slow reaction that can take many years, but the combustion of butane in a fire is a reaction that takes place in fractions of a second. Chemical kinetics is the part of physical chemistry that studies reaction rates.
Warfarin	Warfarin is an anticoagulant normally used in the prevention of thrombosis and thromboembolism, the formation of blood clots in the blood vessels and their migration elsewhere in the body respectively. It was initially introduced in 1948 as a pesticide against rats and mice and is still used for this purpose, although more potent poisons such as brodifacoum have since been developed. In the early 1950s warfarin was found to be effective and relatively safe for preventing thrombosis and embolism (abnormal formation and migration of blood clots) in many disorders.
Order	Order (subtitled 'A Journal on the Theory of Ordered Sets and its Applications') is a quarterly peer-reviewed academic journal on order theory and its applications, published by Springer Science+Business Media. It was founded in 1984 by University of Calgary mathematics professor Ivan Rival; as of 2010, its editor in chief is Dwight Duffus, the Goodrich C. White Professor of Mathematics ' Computer Science at Emory University and a former student of Rival's. According to the Journal Citation Reports, the 2009 impact factor of Order is 0.408, placing it in the fourth quartile of ranked mathematics journals.
Order of reaction	In chemical kinetics, the order of reaction with respect to certain reactant is defined as the index to which its concentration term in the rate equation is raised.

Chapter 14. Kinetics: what affects the speed of a reaction?

For example, given a chemical reaction 2A + B → C with a rate equation

$r = k[A]^2[B]^1$

The reaction order with respect to A in this case is 2 and with respect to B in this case is 1; the overall reaction order is 2 + 1 = 3. It is not necessary that the order of a reaction be a whole number - zero and fractional values of order are possible - but they tend to be integers. Reaction orders can be determined only by experiment.

Collision theory	Collision theory is a theory proposed by William Lewis in 1916 and 1918, that qualitatively explains how chemical reactions occur and why reaction rates differ for different reactions. The collision theory can only occur when the suitable particles of the reactant hit with each other. Only a certain percentage of the sum of the collisions cause any noticeable or significant chemical change; these successful changes are called successful collisions.
Avogadro constant	In chemistry and physics, the Avogadro constant is defined as the ratio of the number of constituent particles N (usually atoms or molecules) in a sample to the amount of substance n (unit mole) through the relationship $N_A = Nn$. Thus, it is the proportionality factor that relates the molar mass of an entity, i.e., the mass per amount of substance, to the mass of said entity. The Avogadro constant expresses the number of elementary entities per mole of substance and it has the value 6.022 141 29(27) × 10^{23} mol^{-1}.
Pasteurization	Pasteurization is a process of heating a food, usually a liquid, to a specific temperature for a definite length of time and then cooling it immediately. This process slows spoilage due to microbial growth in the food. Unlike sterilization, pasteurization is not intended to kill all micro-organisms in the food.
Activation energy	In chemistry, activation energy is a term introduced in 1889 by the Swedish scientist Svante Arrhenius that is defined as the energy that must be overcome in order for a chemical reaction to occur. Activation energy may also be defined as the minimum energy required to start a chemical reaction. The activation energy of a reaction is usually denoted by E_a, and given in units of kilojoules per mole.

Chapter 14. Kinetics: what affects the speed of a reaction?

Transition state	The transition state of a chemical reaction is a particular configuration along the reaction coordinate. It is defined as the state corresponding to the highest energy along this reaction coordinate. At this point, assuming a perfectly irreversible reaction, colliding reactant molecules will always go on to form products.
Catalysis	Catalysis is the change in rate of a chemical reaction due to the participation of a substance called a catalyst. Unlike other reagents that participate in the chemical reaction, a catalyst is not consumed by the reaction itself. A catalyst may participate in multiple chemical transformations.
Catalytic converter	A catalytic converter is an exhaust emission control device which converts toxic chemicals in the exhaust of an internal combustion engine into less toxic substances. Inside a catalytic converter, a catalyst stimulates a chemical reaction in which toxic byproducts of combustion are converted to less toxic substances by way of catalysed chemical reactions. The specific reactions vary with the type of catalyst installed.
Enzyme	Enzymes ( ) are biological molecules that catalyze (i.e., increase the rates of) chemical reactions. In enzymatic reactions, the molecules at the beginning of the process, called substrates, are converted into different molecules, called products. Almost all chemical reactions in a biological cell need enzymes in order to occur at rates sufficient for life.
Ribozyme	A ribozyme is an RNA molecule with a well defined tertiary structure that enables it to perform a chemical reaction. Many ribozymes are catalytic, but some such as self-cleaving ribozymes are consumed by their reactions. Ribozyme means ribonucleic acid enzyme.
Active site	In biology the active site is part of an enzyme where substrates bind and undergo a chemical reaction. The majority of enzymes are proteins but RNA enzymes called ribozymes also exist. The active site of an enzyme is usually found in a cleft or pocket that is lined by amino acid residues that participate in recognition of the substrate.
Catalase	Catalase is a common enzyme found in nearly all living organisms exposed to oxygen. It catalyzes the decomposition of hydrogen peroxide to water and oxygen. It is a very important enzyme in reproductive reactions.
Enzyme catalysis	Enzyme catalysis is the catalysis of chemical reactions by specialized proteins known as enzymes. Catalysis of biochemical reactions in the cell is vital due to the very low reaction rates of the uncatalysed reactions.

Chapter 14. Kinetics: what affects the speed of a reaction?

	The mechanism of enzyme catalysis is similar in principle to other types of chemical catalysis.
Enzyme kinetics	Enzyme kinetics is the study of the chemical reactions that are catalysed by enzymes. In enzyme kinetics, the reaction rate is measured and the effects of varying the conditions of the reaction investigated. Studying an enzyme's kinetics in this way can reveal the catalytic mechanism of this enzyme, its role in metabolism, how its activity is controlled, and how a drug or an agonist might inhibit the enzyme.
Hexokinase	A hexokinase is an enzyme that phosphorylates a six-carbon sugar, a hexose, to a hexose phosphate. In most tissues and organisms, glucose is the most important substrate of hexokinases, and glucose-6-phosphate the most important product. Variation across species Genes that encode hexokinase have been discovered in each domain of life, ranging from bacteria, yeast, and plants to humans and other vertebrates.
Phenylketonuria	Phenylketonuria is an autosomal recessive metabolic genetic disorder characterized by an error in the genetic code for the hepatic enzyme phenylalanine hydroxylase (PAH), rendering it nonfunctional. This enzyme is necessary to metabolize the amino acid phenylalanine (Phe) to the amino acid tyrosine. When PAH enzymatic activity is reduced, phenylalanine accumulates and is converted into phenylpyruvate (also known as phenylketone), which is detected in the urine.
Electron pair	In chemistry, an electron pair consists of two electrons that occupy the same orbital but have opposite spins. Because electrons are fermions, the Pauli exclusion principle forbids these particles from having exactly the same quantum numbers. Therefore the only way to occupy the same orbital, i.e. have the same orbital quantum numbers, is to differ in the spin quantum number.

Chapter 14. Kinetics: what affects the speed of a reaction?

Turnover number	Turnover number has two related meanings:
	In enzymology, turnover number is defined as the maximum number of molecules of substrate that an enzyme can convert to product per catalytic site per unit of time and can be calculated as follows: $k_{cat} = V_{max}/[E]_T$. For example, carbonic anhydrase has a turnover number of 400,000 to 600,000 s^{-1}, which means that each carbonic anhydrase molecule can produce up to 600,000 molecules of product (bicarbonate ions) per second.
	In other chemical fields, such as organometallic catalysis, turnover number is used with a slightly different meaning: the number of moles of substrate that a mole of catalyst can convert before becoming inactivated.

PRACTICE QUIZ
Chapter 14. Kinetics: what affects the speed of a reaction?

1. _____ is a common enzyme found in nearly all living organisms exposed to oxygen. It catalyzes the decomposition of hydrogen peroxide to water and oxygen. It is a very important enzyme in reproductive reactions.

    a. Catechol dioxygenase
    b. Catechol oxidase
    c. Catalase
    d. Cellulase

2. In chemistry, a _____, is one that has equal amounts of left- and right-handed enantiomers of a chiral molecule. The first known _____ was 'racemic acid', which Louis Pasteur found to be a mixture of the two enantiomeric isomers of tartaric acid.

    Nomenclature
    A _____ is denoted by the prefix (±)- or dl- (for sugars the prefix DL- may be used), indicating an equal (1:1) mixture of dextro and levo isomers.

    a. Racemization
    b. Ray-Dutt twist
    c. Regioselectivity
    d. Racemic mixture

3. _____ has two related meanings:

    In enzymology, _____ is defined as the maximum number of molecules of substrate that an enzyme can convert to product per catalytic site per unit of time and can be calculated as follows: $k_{cat} = V_{max}/[E]_T$. For example, carbonic anhydrase has a _____ of 400,000 to 600,000 s^{-1}, which means that each carbonic anhydrase molecule can produce up to 600,000 molecules of product (bicarbonate ions) per second.

    In other chemical fields, such as organometallic catalysis, _____ is used with a slightly different meaning: the number of moles of substrate that a mole of catalyst can convert before becoming inactivated.

---

a. Variational Transition State Theory
b. Turnover number
c. KinITC
d. Law of mass action

4. _____s are proteins that are composed of DNA-binding domains and thus have a specific or general affinity for either single or double stranded DNA. Sequence-specific _____s generally interact with the major groove of B-DNA, because it exposes more functional groups that identify a base pair. However there are some known minor groove DNA-binding ligands such as Netropsin, Distamycin, Hoechst 33258, Pentamidine and others.

Examples
_____s include transcription factors which modulate the process of transcription, various polymerases, nucleases which cleave DNA molecules, and histones which are involved in chromosome packaging and transcription in the cell nucleus.

a. Flory convention
b. Fluorescence recovery after photobleaching
c. FlyFactorSurvey
d. DNA-binding protein

5. A _____ is an RNA molecule with a well defined tertiary structure that enables it to perform a chemical reaction. Many _____s are catalytic, but some such as self-cleaving _____s are consumed by their reactions. _____ means ribonucleic acid enzyme.

a. SAR1A
b. Siderophore
c. Sphingosine
d. Ribozyme

ANSWER KEY
Chapter 14. Kinetics: what affects the speed of a reaction?

1. c
2. d
3. b
4. d
5. d

You can take the complete Chapter Practice Test

for **Chapter 14. Kinetics: what affects the speed of a reaction?**
on all key terms, persons, places, and concepts.

Online 99 Cents

http://www.epub14.51.19910.14.cram101.com/

Use www.Cram101.com for all your study needs

including Cram101's online interactive problem solving labs in chemistry, statistics, mathematics, and more.

CHAPTER OUTLINE: KEY TERMS, PEOPLE, PLACES, CONCEPTS
Chapter 15
Equilibria: how far do reactions go?

- Avogadro constant
- Melting
- Racemic mixture
- DNA profiling
- Dynamic equilibrium
- Equilibrium constant
- Nitrogen
- Reaction quotient
- Dissociation constant
- Lactate dehydrogenase
- Lactic acid
- Osmosis
- Osmotic pressure
- Intramolecular force
- Partial pressure
- DNA-binding protein
- X-ray crystallography
- Gas constant
- Metabolism

CHAPTER HIGHLIGHTS: KEY TERMS, PEOPLE, PLACES, CONCEPTS
Chapter 15. Equilibria: how far do reactions go?

Avogadro constant	In chemistry and physics, the Avogadro constant is defined as the ratio of the number of constituent particles N (usually atoms or molecules) in a sample to the amount of substance n (unit mole) through the relationship $N_A = Nn$. Thus, it is the proportionality factor that relates the molar mass of an entity, i.e., the mass per amount of substance, to the mass of said entity. The Avogadro constant expresses the number of elementary entities per mole of substance and it has the value $6.022\ 141\ 29(27) \times 10^{23}$ mol^{-1}.
Melting	Melting, is a physical process that results in the phase transition of a substance from a solid to a liquid. The internal energy of a substance is increased, typically by the application of heat or pressure, resulting in a rise of its temperature to the melting point, at which the rigid ordering of molecular entities in the solid breaks down to a less-ordered state and the solid liquefies. An object that has melted completely is molten.
Racemic mixture	In chemistry, a racemic mixture, is one that has equal amounts of left- and right-handed enantiomers of a chiral molecule. The first known racemic mixture was 'racemic acid', which Louis Pasteur found to be a mixture of the two enantiomeric isomers of tartaric acid. Nomenclature A racemic mixture is denoted by the prefix (±)- or dl- (for sugars the prefix DL- may be used), indicating an equal (1:1) mixture of dextro and levo isomers.
DNA profiling	DNA profiling is a technique employed by forensic scientists to assist in the identification of individuals by their respective DNA profiles. DNA profiles are encrypted sets of numbers that reflect a person's DNA makeup, which can also be used as the person's identifier. DNA profiling should not be confused with full genome sequencing.
Dynamic equilibrium	A dynamic equilibrium exists once a reversible reaction ceases to change its ratio of reactants/products, but substances move between the chemicals at an equal rate, meaning there is no net change. It is a particular example of a system in a steady state. In thermodynamics a closed system is in thermodynamic equilibrium when reactions occur at such rates that the composition of the mixture does not change with time.
Equilibrium constant	For a general chemical equilibrium $$\alpha A + \beta B ... \rightleftharpoons \sigma S + \tau T ...$$

Chapter 15. Equilibria: how far do reactions go?

the equilibrium constant can be defined such that, at equilibrium,

$$K = \frac{\{S\}^\sigma \{T\}^\tau \ldots}{\{A\}^\alpha \{B\}^\beta \ldots}$$

where {A} is the activity of the chemical species A, etc. (activity is a dimensionless quantity). It is conventional to put the activities of the products in the numerator and those of the reactants in the denominator.

Nitrogen

Nitrogen is a chemical element that has the symbol N, atomic number of 7 and atomic mass 14.00674 u. Elemental nitrogen is a colorless, odorless, tasteless, and mostly inert diatomic gas at standard conditions, constituting 78.09% by volume of Earth's atmosphere. The element nitrogen was discovered as a separable component of air, by Scottish physician Daniel Rutherford, in 1772.

Reaction quotient

In chemistry, a reaction quotient: Q_r is a function of the activities or concentrations of the chemical species involved in a chemical reaction. In the special case that the reaction is at equilibrium the reaction quotient is equal to the equilibrium constant.

A general chemical reaction in which α moles of a reactant A and β moles of a reactant B react to give σ moles of a product S and τ moles of a product T can be written as

αA + βB σS + τT

The reaction is written as an equilibrium even though in many cases it may appear to have gone to completion.

Chapter 15. Equilibria: how far do reactions go?

Dissociation constant	In chemistry, biochemistry, and pharmacology, a dissociation constant is a specific type of equilibrium constant that measures the propensity of a larger object to separate (dissociate) reversibly into smaller components, as when a complex falls apart into its component molecules, or when a salt splits up into its component ions. The dissociation constant is usually denoted K_d and is the inverse of the association constant. In the special case of salts, the dissociation constant can also be called an ionization constant.
Lactate dehydrogenase	D-lactate dehydrogenase, membrane binding Lactate dehydrogenase is an enzyme (EC 1.1.1.27) present in a wide variety of organisms, including plants and animals. Lactate dehydrogenases exist in four distinct enzyme classes. Two of them are cytochrome c-dependent enzymes, each acting on either D-lactate (EC 1.1.2.4) or L-lactate (EC 1.1.2.3).
Lactic acid	Lactic acid, is a chemical compound that plays a role in various biochemical processes and was first isolated in 1780 by the Swedish chemist Carl Wilhelm Scheele. Lactic acid is a carboxylic acid with the chemical formula $C_3H_6O_3$. It has a hydroxyl group adjacent to the carboxyl group, making it an alpha hydroxy acid (AHA).
Osmosis	Osmosis is the net movement of solvent molecules through a partially permeable membrane into a region of higher solute concentration, in order to equalize the solute concentrations on the two sides. It may also be used to describe a physical process in which any solvent moves, without input of energy, across a semipermeable membrane (permeable to the solvent, but not the solute) separating two solutions of different concentrations. Although osmosis does not require input of energy, it does use kinetic energy and can be made to do work.
Osmotic pressure	Osmotic pressure is the pressure which needs to be applied to a solution to prevent the inward flow of water across a semipermeable membrane. It is also defined as the minimum pressure needed to nullify osmosis.

Chapter 15. Equilibria: how far do reactions go?

	The phenomenon of osmotic pressure arises from the tendency of a pure solvent to move through a semi-permeable membrane and into a solution containing a solute to which the membrane is impermeable.
Intramolecular force	An intramolecular force is any force that holds together the atoms making up a molecule or compound. They contain all types of chemical bond. They are stronger than intermolecular forces, which are present between atoms or molecules that are not actually bonded.
Partial pressure	In a mixture of ideal gases, each gas has a partial pressure which is the pressure which the gas would have if it alone occupied the volume. The total pressure of a gas mixture is the sum of the partial pressures of each individual gas in the mixture.
	In chemistry, the partial pressure of a gas in a mixture of gases is defined as above.
DNA-binding protein	DNA-binding proteins are proteins that are composed of DNA-binding domains and thus have a specific or general affinity for either single or double stranded DNA. Sequence-specific DNA-binding proteins generally interact with the major groove of B-DNA, because it exposes more functional groups that identify a base pair. However there are some known minor groove DNA-binding ligands such as Netropsin, Distamycin, Hoechst 33258, Pentamidine and others. Examples DNA-binding proteins include transcription factors which modulate the process of transcription, various polymerases, nucleases which cleave DNA molecules, and histones which are involved in chromosome packaging and transcription in the cell nucleus.
X-ray crystallography	X-ray crystallography is a method of determining the arrangement of atoms within a crystal, in which a beam of X-rays strikes a crystal and causes the beam of light to spread into many specific directions. From the angles and intensities of these diffracted beams, a crystallographer can produce a three-dimensional picture of the density of electrons within the crystal. From this electron density, the mean positions of the atoms in the crystal can be determined, as well as their chemical bonds, their disorder and various other information.

Chapter 15. Equilibria: how far do reactions go?

Gas constant	The gas constant is a physical constant which is featured in many fundamental equations in the physical sciences, such as the ideal gas law and the Nernst equation. It is equivalent to the Boltzmann constant, but expressed in units of energy (i.e. the pressure-volume product) per temperature increment per mole (rather than energy per temperature increment per particle). The constant is also a combination of the constants from Boyle's Law, Charles' Law, Avogadro's Law, and Gay-Lussac's Law.
Metabolism	Metabolism is the set of chemical reactions that happen in the cells of living organisms to sustain life. These processes allow organisms to grow and reproduce, maintain their structures, and respond to their environments. The word metabolism can also refer to all chemical reactions that occur in living organisms, including digestion and the transport of substances into and between different cells, in which case the set of reactions within the cells is called intermediary metabolism or intermediate metabolism.

PRACTICE QUIZ
Chapter 15. Equilibria: how far do reactions go?

1. _____, is a physical process that results in the phase transition of a substance from a solid to a liquid. The internal energy of a substance is increased, typically by the application of heat or pressure, resulting in a rise of its temperature to the _____ point, at which the rigid ordering of molecular entities in the solid breaks down to a less-ordered state and the solid liquefies. An object that has melted completely is molten.

    a. Melting
    b. Melting-point depression
    c. Mesophase
    d. Mpemba effect

2. In chemistry and physics, the _____ is defined as the ratio of the number of constituent particles N (usually atoms or molecules) in a sample to the amount of substance n (unit mole) through the relationship N_A = Nn. Thus, it is the proportionality factor that relates the molar mass of an entity, i.e., the mass per amount of substance, to the mass of said entity. The _____ expresses the number of elementary entities per mole of substance and it has the value 6.022 141 29(27) × 10^{23} mol^{-1}.

    a. Avogadro constant
    b. Osmolarity
    c. Osmometer
    d. Osmotic pressure

3. For a general chemical equilibrium

$$\alpha A + \beta B ... \rightleftharpoons \sigma S + \tau T ...$$

   the _____ can be defined such that, at equilibrium,

$$K = \frac{\{S\}^\sigma \{T\}^\tau ...}{\{A\}^\alpha \{B\}^\beta ...}$$

   where {A} is the activity of the chemical species A, etc. (activity is a dimensionless quantity). It is conventional to put the activities of the products in the numerator and those of the reactants in the denominator.

---

a. Equilibrium unfolding
b. ICE table
c. Ion-association
d. Equilibrium constant

4. D-_____, membrane binding

   _____ is an enzyme (EC 1.1.1.27) present in a wide variety of organisms, including plants and animals.

   _____s exist in four distinct enzyme classes. Two of them are cytochrome c-dependent enzymes, each acting on either D-lactate (EC 1.1.2.4) or L-lactate (EC 1.1.2.3).

   a. Lactate dehydrogenase
   b. Medullary thyroid cancer
   c. Neprilysin
   d. NMP22

5. In chemistry, biochemistry, and pharmacology, a _____ is a specific type of equilibrium constant that measures the propensity of a larger object to separate (dissociate) reversibly into smaller components, as when a complex falls apart into its component molecules, or when a salt splits up into its component ions. The _____ is usually denoted K_d and is the inverse of the association constant. In the special case of salts, the _____ can also be called an ionization constant.

   a. Dissociation constant
   b. Dynamic equilibrium
   c. Hand boiler
   d. Hemoglobin

ANSWER KEY
Chapter 15. Equilibria: how far do reactions go?

1. a
2. a
3. d
4. a
5. a

You can take the complete Chapter Practice Test

for Chapter 15. Equilibria: how far do reactions go?
on all key terms, persons, places, and concepts.

Online 99 Cents

http://www.epub14.51.19910.15.cram101.com/

Use www.Cram101.com for all your study needs

including Cram101's online interactive problem solving labs in chemistry, statistics, mathematics, and more.

CHAPTER OUTLINE: KEY TERMS, PEOPLE, PLACES, CONCEPTS
Chapter 16
Acids, bases, and the aqueous environment: the medium of life

- Avogadro constant
- Hydrogen chloride
- Hydrolysis
- Racemic mixture
- Carbonic acid
- Nitric acid
- Strong acid
- Weak acid
- Acid dissociation constant
- Dissociation constant
- Buffer solution
- Exponential function
- Logarithm
- Natural logarithm
- Active site
- Enzyme catalysis
- Enzyme kinetics
- Intramolecular force
- Concentration

Chapter 16. Acids, bases, and the aqueous environment: the medium of life

- Alkalosis
  - Conjugate acid

CHAPTER HIGHLIGHTS: KEY TERMS, PEOPLE, PLACES, CONCEPTS
Chapter 16. Acids, bases, and the aqueous environment: the medium of life

Avogadro constant	In chemistry and physics, the Avogadro constant is defined as the ratio of the number of constituent particles N (usually atoms or molecules) in a sample to the amount of substance n (unit mole) through the relationship $N_A = Nn$. Thus, it is the proportionality factor that relates the molar mass of an entity, i.e., the mass per amount of substance, to the mass of said entity. The Avogadro constant expresses the number of elementary entities per mole of substance and it has the value $6.022\ 141\ 29(27) \times 10^{23}\ mol^{-1}$.
Hydrogen chloride	The compound hydrogen chloride has the formula HCl. At room temperature, it is a colorless gas, which forms white fumes of hydrochloric acid upon contact with atmospheric humidity. Hydrogen chloride gas and hydrochloric acid are important in technology and industry.
Hydrolysis	Hydrolysis usually means the rupture of chemical bonds by the addition of water. Generally, hydrolysis is a step in the degradation of a substance. In terms of the word's derivation, hydrolysis comes from Greek roots hydro 'water' + lysis 'separation'.
Racemic mixture	In chemistry, a racemic mixture, is one that has equal amounts of left- and right-handed enantiomers of a chiral molecule. The first known racemic mixture was 'racemic acid', which Louis Pasteur found to be a mixture of the two enantiomeric isomers of tartaric acid. Nomenclature A racemic mixture is denoted by the prefix (±)- or dl- (for sugars the prefix DL- may be used), indicating an equal (1:1) mixture of dextro and levo isomers.
Carbonic acid	Carbonic acid is the organic compound with the formula H_2CO_3 (equivalently $OC(OH)_2$). It is also a name sometimes given to solutions of carbon dioxide in water, because such solutions contain small amounts of H_2CO_3. Carbonic acid forms two kinds of salts, the carbonates and the bicarbonates.
Nitric acid	Nitric acid also known as aqua fortis and spirit of niter, is a highly corrosive and toxic strong mineral acid which is normally colorless but tends to acquire a yellow cast due to the accumulation of oxides of nitrogen if long-stored. Ordinary nitric acid has a concentration of 68%. When the solution contains more than 86% HNO_3, it is referred to as fuming nitric acid.
Strong acid	A strong acid is an acid that ionizes completely in an aqueous solution by losing one proton, according to the equation $$HA(aq) \rightarrow H^+(aq) + A^-(aq)$$

Chapter 16. Acids, bases, and the aqueous environment: the medium of life

For sulfuric acid which is diprotic, the 'strong acid' designation refers only to dissociation of the first proton

$$H_2SO_4(aq) \rightarrow H^+(aq) + HSO_4^-(aq)$$

More precisely, the acid must be stronger in aqueous solution than hydronium ion, so strong acids are acids with a $pK_a < -1.74$. An example is HCl for which $pK_a = -6.3$. This generally means that in aqueous solution at standard temperature and pressure, the concentration of hydronium ions is equal to the concentration of strong acid introduced to the solution. While strong acids are generally assumed to be the most corrosive, this is not always true. The carborane superacid H ($CHB_{11}Cl_{11}$), which is one million times stronger than sulfuric acid, is entirely non-corrosive, whereas the weak acid hydrofluoric acid (HF) is extremely corrosive and can dissolve, among other things, glass and all metals except iridium.

Weak acid	A weak acid is an acid that dissociates incompletely. It does not release all of its hydrogens in a solution, donating only a partial amount of its protons to the solution. These acids have higher pKa than strong acids, which release all of their hydrogen atoms when dissolved in water.
Acid dissociation constant	An acid dissociation constant, K_a, (also known as acidity constant, or acid-ionization constant) is a quantitative measure of the strength of an acid in solution. It is the equilibrium constant for a chemical reaction known as dissociation in the context of acid-base reactions. The equilibrium can be written symbolically as: $$HA \ A^- + H^+,$$ where HA is a generic acid that dissociates by splitting into A^-, known as the conjugate base of the acid, and the hydrogen ion or proton, H^+, which, in the case of aqueous solutions, exists as a solvated hydronium ion.

Chapter 16. Acids, bases, and the aqueous environment: the medium of life

Dissociation constant	In chemistry, biochemistry, and pharmacology, a dissociation constant is a specific type of equilibrium constant that measures the propensity of a larger object to separate (dissociate) reversibly into smaller components, as when a complex falls apart into its component molecules, or when a salt splits up into its component ions. The dissociation constant is usually denoted K_d and is the inverse of the association constant. In the special case of salts, the dissociation constant can also be called an ionization constant.
Buffer solution	A buffer solution is an aqueous solution consisting of a mixture of a weak acid and its conjugate base or a weak base and its conjugate acid. Its pH changes very little when a small amount of strong acid or base is added to it. Buffer solutions are used as a means of keeping pH at a nearly constant value in a wide variety of chemical applications.
Exponential function	In mathematics, the exponential function is the function e^x, where e is the number (approximately 2.718281828) such that the function e^x is its own derivative. The exponential function is used to model a relationship in which a constant change in the independent variable gives the same proportional change (i.e. percentage increase or decrease) in the dependent variable. The function is often written as exp(x), especially when it is impractical to write the independent variable as a superscript.
Logarithm	The logarithm of a number is the exponent by which another fixed value, the base, has to be raised to produce that number. For example, the logarithm of 1000 to base 10 is 3, because 1000 is 10 to the power 3: $1000 = 10^3 = 10?×?10?×?10$. More generally, if $x = b^y$, then y is the logarithm of x to base b, and is written $\log_b(x)$, so $\log_{10}(1000) = 3$. Logarithms were introduced by John Napier in the early 17th century as a means to simplify calculations. They were rapidly adopted by scientists, engineers, and others to perform computations more easily, using slide rules and logarithm tables.
Natural logarithm	The natural logarithm is the logarithm to the base e, where e is an irrational and transcendental constant approximately equal to 2.718 281 828. The natural logarithm is generally written as ln(x), $\log_e(x)$ or sometimes, if the base of e is implicit, as simply log(x). The natural logarithm of a number x is the power to which e would have to be raised to equal x.

Chapter 16. Acids, bases, and the aqueous environment: the medium of life

Active site	In biology the active site is part of an enzyme where substrates bind and undergo a chemical reaction. The majority of enzymes are proteins but RNA enzymes called ribozymes also exist. The active site of an enzyme is usually found in a cleft or pocket that is lined by amino acid residues that participate in recognition of the substrate.
Enzyme catalysis	Enzyme catalysis is the catalysis of chemical reactions by specialized proteins known as enzymes. Catalysis of biochemical reactions in the cell is vital due to the very low reaction rates of the uncatalysed reactions. The mechanism of enzyme catalysis is similar in principle to other types of chemical catalysis.
Enzyme kinetics	Enzyme kinetics is the study of the chemical reactions that are catalysed by enzymes. In enzyme kinetics, the reaction rate is measured and the effects of varying the conditions of the reaction investigated. Studying an enzyme's kinetics in this way can reveal the catalytic mechanism of this enzyme, its role in metabolism, how its activity is controlled, and how a drug or an agonist might inhibit the enzyme.
Intramolecular force	An intramolecular force is any force that holds together the atoms making up a molecule or compound. They contain all types of chemical bond. They are stronger than intermolecular forces, which are present between atoms or molecules that are not actually bonded.
Concentration	In chemistry, concentration is defined as the abundance of a constituent divided by the total volume of a mixture. Furthermore, in chemistry, four types of mathematical description can be distinguished: mass concentration, molar concentration, number concentration, and volume concentration. The term concentration can be applied to any kind of chemical mixture, but most frequently it refers to solutes in solutions.
Alkalosis	Alkalosis refers to a condition reducing hydrogen ion concentration of arterial blood plasma (alkalemia). Generally, alkalosis is said to occur when pH of the blood exceeds 7.45. The opposite condition is acidosis. Types

Chapter 16. Acids, bases, and the aqueous environment: the medium of life

Alkalosis can refer to:

- Respiratory alkalosis
- Metabolic alkalosis

The main cause of respiratory alkalosis is hyperventilation, resulting in a loss of carbon dioxide.

Conjugate acid	Within the Brønsted-Lowry acid-base theory (protonic), a conjugate acid is the acid member, HX, of a pair of two compounds that transform into each other by gain or loss of a proton (hydrogen ion). A conjugate acid can also be seen as the chemical substance that releases, or donates, a proton (hydrogen ion) in the forward chemical reaction, hence, the term acid. The base produced, X^-, is called the conjugate base, and it absorbs, or gains, a proton in the backward chemical reaction.

PRACTICE QUIZ
Chapter 16. Acids, bases, and the aqueous environment: the medium of life

1. _____ usually means the rupture of chemical bonds by the addition of water. Generally, _____ is a step in the degradation of a substance. In terms of the word's derivation, _____ comes from Greek roots hydro 'water' + lysis 'separation'.

    a. Jacobi coordinates
    b. Liesegang rings
    c. Hydrolysis
    d. Pyrophorus

2. _____ is the study of the chemical reactions that are catalysed by enzymes. In _____, the reaction rate is measured and the effects of varying the conditions of the reaction investigated. Studying an enzyme's kinetics in this way can reveal the catalytic mechanism of this enzyme, its role in metabolism, how its activity is controlled, and how a drug or an agonist might inhibit the enzyme.

    a. 1,2-Dioxetanedione
    b. Ada
    c. ADAM17
    d. Enzyme kinetics

3. _____ refers to a condition reducing hydrogen ion concentration of arterial blood plasma (alkalemia). Generally, _____ is said to occur when pH of the blood exceeds 7.45. The opposite condition is acidosis.

    Types

    _____ can refer to:

    - Respiratory _____
    - Metabolic _____

    The main cause of respiratory _____ is hyperventilation, resulting in a loss of carbon dioxide.

    a. 1,2-Dioxetanedione
    b. Congruent number
    c. Alkalosis
    d. Covering set

---

4. In chemistry and physics, the _____ is defined as the ratio of the number of constituent particles N (usually atoms or molecules) in a sample to the amount of substance n (unit mole) through the relationship $N_A = Nn$. Thus, it is the proportionality factor that relates the molar mass of an entity, i.e., the mass per amount of substance, to the mass of said entity. The _____ expresses the number of elementary entities per mole of substance and it has the value $6.022\ 141\ 29(27) \times 10^{23}\ mol^{-1}$.

   a. Equivalent weight
   b. Osmolarity
   c. Osmometer
   d. Avogadro constant

5. A _____ is an aqueous solution consisting of a mixture of a weak acid and its conjugate base or a weak base and its conjugate acid. Its pH changes very little when a small amount of strong acid or base is added to it. _____s are used as a means of keeping pH at a nearly constant value in a wide variety of chemical applications.

   a. Buffering agent
   b. Carbonate alkalinity
   c. Buffer solution
   d. Chemical field-effect transistor

ANSWER KEY
Chapter 16. Acids, bases, and the aqueous environment: the medium of life

1. c
2. d
3. c
4. d
5. c

You can take the complete Chapter Practice Test

for Chapter 16. Acids, bases, and the aqueous environment: the medium of life

on all key terms, persons, places, and concepts.

Online 99 Cents

http://www.epub14.51.19910.16.cram101.com/

Use www.Cram101.com for all your study needs

including Cram101's online interactive problem solving labs in chemistry, statistics, mathematics, and more.

CHAPTER OUTLINE: KEY TERMS, PEOPLE, PLACES, CONCEPTS
Chapter 17
Chemical reactions 1: bringing molecules to life

- Chemical reaction
- Glycolysis
- Racemic mixture
- Reaction rate
- Stoichiometry
- Triple bond
- Reaction mechanism
- Electrophile
- Nucleophile
- DNA profiling
- Hydrogen bond
- Hydrogen bromide
- Bond cleavage
- Enzyme catalysis
- Malate dehydrogenase
- Anaerobic respiration
- Lactic acid
- Metabolism
- X-ray crystallography

Chapter 17. Chemical reactions 1: bringing molecules to life

_____ Hydrogen peroxide

_____ Hydroxyl radical

_____ Reactive oxygen species

_____ Chain reaction

_____ Ozone

_____ Vitamin E

_____ Antioxidant

_____ Oxygen toxicity

CHAPTER HIGHLIGHTS: KEY TERMS, PEOPLE, PLACES, CONCEPTS
Chapter 17. Chemical reactions 1: bringing molecules to life

Chemical reaction	A chemical reaction is a process that leads to the transformation of one set of chemical substances to another. Chemical reactions can be either spontaneous, requiring no input of energy, or non-spontaneous, typically following the input of some type of energy, such as heat, light or electricity. Classically, chemical reactions encompass changes that strictly involve the motion of electrons in the forming and breaking of chemical bonds, although the general concept of a chemical reaction, in particular the notion of a chemical equation, is applicable to transformations of elementary particles (such as illustrated by Feynman diagrams), as well as nuclear reactions.
Glycolysis	Glycolysis is the metabolic pathway that converts glucose $C_6H_{12}O_6$, into pyruvate, $CH_3COCOO^- + H^+$. The free energy released in this process is used to form the high-energy compounds ATP (adenosine triphosphate), $FADH_2$ and NADH (reduced nicotinamide adenine dinucleotide). Glycolysis is a definite sequence of ten reactions involving ten intermediate compounds (one of the steps involves two intermediates).
Racemic mixture	In chemistry, a racemic mixture, is one that has equal amounts of left- and right-handed enantiomers of a chiral molecule. The first known racemic mixture was 'racemic acid', which Louis Pasteur found to be a mixture of the two enantiomeric isomers of tartaric acid. Nomenclature A racemic mixture is denoted by the prefix (±)- or dl- (for sugars the prefix DL- may be used), indicating an equal (1:1) mixture of dextro and levo isomers.
Reaction rate	The reaction rate or speed of reaction for a reactant or product in a particular reaction is intuitively defined as how fast or slow a reaction takes place. For example, the oxidation of iron under the atmosphere is a slow reaction that can take many years, but the combustion of butane in a fire is a reaction that takes place in fractions of a second. Chemical kinetics is the part of physical chemistry that studies reaction rates.

Chapter 17. Chemical reactions 1: bringing molecules to life

Stoichiometry

Stoichiometry is a branch of chemistry that deals with the relative quantities of reactants and products in chemical reactions. In a balanced chemical reaction, the relations among quantities of reactants and products typically form a ratio of whole numbers. For example, in a reaction that forms ammonia (NH_3), exactly one molecule of nitrogen (N_2) reacts with three molecules of hydrogen (H_2) to produce two molecules of NH_3:

$$N_2 + 3 H_2 \rightarrow 2 NH_3$$

Stoichiometry can be used to find quantities such as the amount of products (in mass, moles, volume, etc).

Triple bond

A triple bond in chemistry is a chemical bond between two chemical elements involving six bonding electrons instead of the usual two in a covalent single bond. The most common triple bond, that between two carbon atoms, can be found in alkynes. Other functional groups containing a triple bond are cyanides and isocyanides.

Reaction mechanism

In chemistry, a reaction mechanism is the step by step sequence of elementary reactions by which overall chemical change occurs.

Although only the net chemical change is directly observable for most chemical reactions, experiments can often be designed that suggest the possible sequence of steps in a reaction mechanism. Recently, electrospray ionization mass spectrometry has been used to corroborate the mechanism of several organic reaction proposals.

Electrophile

In general, electrophiles are positively charged species that are attracted to an electron rich centre. In chemistry, an electrophile is a reagent attracted to electrons that participates in a chemical reaction by accepting an electron pair in order to bond to a nucleophile. Because electrophiles accept electrons, they are Lewis acids .

Nucleophile

A nucleophile is a species that donates an electron-pair to an electrophile to form a chemical bond in a reaction. All molecules or ions with a free pair of electrons or at least one pi bond can act as nucleophiles. Because nucleophiles donate electrons, they are by definition Lewis bases.

Chapter 17. Chemical reactions 1: bringing molecules to life

DNA profiling	DNA profiling is a technique employed by forensic scientists to assist in the identification of individuals by their respective DNA profiles. DNA profiles are encrypted sets of numbers that reflect a person's DNA makeup, which can also be used as the person's identifier. DNA profiling should not be confused with full genome sequencing.
Hydrogen bond	A hydrogen bond is the attractive interaction of a hydrogen atom with an electronegative atom, such as nitrogen, oxygen or fluorine, that comes from another molecule or chemical group. The hydrogen has a polar bonding to another electronegative atom to create the bond. These bonds can occur between molecules (intermolecularly), or within different parts of a single molecule (intramolecularly).
Hydrogen bromide	Hydrogen bromide is the diatomic molecule HBr. HBr is a gas at standard conditions. Hydrobromic acid forms upon dissolving HBr in water.
Bond cleavage	Bond cleavage, is the splitting of chemical bonds. If the two electrons in a cleaved covalent bond are divided between the products, the process is known as homolytic fission or homolysis and free redicals are generated by homolytic cleavage. Alternatively, the case where both electrons are retained by one product and charged species that is nucleophile and electrophile are generated by the process is known as heterolytic fission and (heterolysis).
Enzyme catalysis	Enzyme catalysis is the catalysis of chemical reactions by specialized proteins known as enzymes. Catalysis of biochemical reactions in the cell is vital due to the very low reaction rates of the uncatalysed reactions. The mechanism of enzyme catalysis is similar in principle to other types of chemical catalysis.
Malate dehydrogenase	Malate dehydrogenase (MDH) is an enzyme that reversibly catalyzes the oxidation of malate to oxaloacetate using the reduction of NAD+ to NADH. This reaction is part of many metabolic pathways, including the citric acid cycle. Other malate dehydrogenases, which have other EC numbers and catalyze other reactions oxidizing malate, have qualified names like malate dehydrogenase.

Chapter 17. Chemical reactions 1: bringing molecules to life

	Malate dehydrogenase is also involved in gluconeogenesis, the synthesis of glucose from smaller molecules.
Anaerobic respiration	Anaerobic respiration is a form of respiration using electron acceptors other than oxygen. Although oxygen is not used as the final electron acceptor, the process still uses a respiratory electron transport chain; it is respiration without oxygen. In order for the electron transport chain to function, an exogenous final electron acceptor must be present to allow electrons to pass through the system.
Lactic acid	Lactic acid, is a chemical compound that plays a role in various biochemical processes and was first isolated in 1780 by the Swedish chemist Carl Wilhelm Scheele. Lactic acid is a carboxylic acid with the chemical formula $C_3H_6O_3$. It has a hydroxyl group adjacent to the carboxyl group, making it an alpha hydroxy acid (AHA).
Metabolism	Metabolism is the set of chemical reactions that happen in the cells of living organisms to sustain life. These processes allow organisms to grow and reproduce, maintain their structures, and respond to their environments. The word metabolism can also refer to all chemical reactions that occur in living organisms, including digestion and the transport of substances into and between different cells, in which case the set of reactions within the cells is called intermediary metabolism or intermediate metabolism.
X-ray crystallography	X-ray crystallography is a method of determining the arrangement of atoms within a crystal, in which a beam of X-rays strikes a crystal and causes the beam of light to spread into many specific directions. From the angles and intensities of these diffracted beams, a crystallographer can produce a three-dimensional picture of the density of electrons within the crystal. From this electron density, the mean positions of the atoms in the crystal can be determined, as well as their chemical bonds, their disorder and various other information.
Hydrogen peroxide	Hydrogen peroxide is the simplest peroxide (a compound with an oxygen-oxygen single bond). It is also a strong oxidizer. Hydrogen peroxide is a clear liquid, slightly more viscous than water.
Hydroxyl radical	The hydroxyl radical, •OH, is the neutral form of the hydroxide ion (OH^-). Hydroxyl radicals are highly reactive and consequently short-lived; however, they form an important part of radical chemistry. Most notably hydroxyl radicals are produced from the decomposition of hydroperoxides (ROOH) or, in atmospheric chemistry, by the reaction of excited atomic oxygen with water.

Chapter 17. Chemical reactions 1: bringing molecules to life

Reactive oxygen species	Reactive oxygen species are chemically reactive molecules containing oxygen. Examples include oxygen ions and peroxides. Reactive oxygen species are highly reactive due to the presence of unpaired valence shell electrons.
Chain reaction	A chain reaction is a sequence of reactions where a reactive product or by-product causes additional reactions to take place. In a chain reaction, positive feedback leads to a self-amplifying chain of events.
	Chain reactions are one way in which systems which are in thermodynamic non-equilibrium can release energy or increase entropy in order to reach a state of higher entropy.
Ozone	Ozone or trioxygen, is a triatomic molecule, consisting of three oxygen atoms. It is an allotrope of oxygen that is much less stable than the diatomic allotrope (O_2), breaking down with a half life of about half an hour in the lower atmosphere, to normal dioxygen. Ozone is formed from dioxygen by the action of ultraviolet light and also atmospheric electrical discharges, and is present in low concentrations throughout the Earth's atmosphere.
Vitamin E	Vitamin E is a generic term for tocopherols which taken from the Greek words tokos, meaning offspring, and phero, meaning to bear, and tocotrienols. Vitamin E is a family of α-, β-, γ-, and δ- (respectively: alpha, beta, gamma, and delta) tocopherols and corresponding four tocotrienols. Vitamin E is a fat-soluble antioxidant that stops the production of reactive oxygen species formed when fat undergoes oxidation.
Antioxidant	An antioxidant is a molecule that inhibits the oxidation of other molecules. Oxidation is a chemical reaction that transfers electrons or hydrogen from a substance to an oxidizing agent. Oxidation reactions can produce free radicals.
Oxygen toxicity	Oxygen toxicity is a condition resulting from the harmful effects of breathing molecular oxygen (O_2) at elevated partial pressures. It is also known as oxygen toxicity syndrome, oxygen intoxication, and oxygen poisoning. Historically, the central nervous system condition was called the Paul Bert effect, and the pulmonary condition the Lorrain Smith effect, after the researchers who pioneered its discovery and description in the late 19th century.

PRACTICE QUIZ
Chapter 17. Chemical reactions 1: bringing molecules to life

1. _____ is a condition resulting from the harmful effects of breathing molecular oxygen (O_2) at elevated partial pressures. It is also known as _____ syndrome, oxygen intoxication, and oxygen poisoning. Historically, the central nervous system condition was called the Paul Bert effect, and the pulmonary condition the Lorrain Smith effect, after the researchers who pioneered its discovery and description in the late 19th century.

    a. Oxygenation
    b. Ozone layer
    c. Ozone-oxygen cycle
    d. Oxygen toxicity

2. In chemistry, a _____, is one that has equal amounts of left- and right-handed enantiomers of a chiral molecule. The first known _____ was 'racemic acid', which Louis Pasteur found to be a mixture of the two enantiomeric isomers of tartaric acid.

    Nomenclature
    A _____ is denoted by the prefix (±)- or dl- (for sugars the prefix DL- may be used), indicating an equal (1:1) mixture of dextro and levo isomers.

    a. Racemization
    b. Ray-Dutt twist
    c. Regioselectivity
    d. Racemic mixture

3. A _____ is the attractive interaction of a hydrogen atom with an electronegative atom, such as nitrogen, oxygen or fluorine, that comes from another molecule or chemical group. The hydrogen has a polar bonding to another electronegative atom to create the bond. These bonds can occur between molecules (intermolecularly), or within different parts of a single molecule (intramolecularly).

    a. Hydrophobic effect
    b. J-aggregate
    c. Macrocycle
    d. Hydrogen bond

---

4. A _____ is a process that leads to the transformation of one set of chemical substances to another. _____s can be either spontaneous, requiring no input of energy, or non-spontaneous, typically following the input of some type of energy, such as heat, light or electricity. Classically, _____s encompass changes that strictly involve the motion of electrons in the forming and breaking of chemical bonds, although the general concept of a _____, in particular the notion of a chemical equation, is applicable to transformations of elementary particles (such as illustrated by Feynman diagrams), as well as nuclear reactions.

   a. Bradsher cycloaddition
   b. Chemical reaction
   c. Carbothermic reaction
   d. Chemical decomposition

5. _____ is the metabolic pathway that converts glucose $C_6H_{12}O_6$, into pyruvate, $CH_3COCOO^- + H^+$. The free energy released in this process is used to form the high-energy compounds ATP (adenosine triphosphate), $FADH_2$ and NADH (reduced nicotinamide adenine dinucleotide).

   _____ is a definite sequence of ten reactions involving ten intermediate compounds (one of the steps involves two intermediates).

   a. Glycolysis
   b. Melting curve analysis
   c. Metal Ions in Life Sciences
   d. MINAS

ANSWER KEY
Chapter 17. Chemical reactions 1: bringing molecules to life

1. d
2. d
3. d
4. b
5. a

You can take the complete Chapter Practice Test

for Chapter 17. Chemical reactions 1: bringing molecules to life
on all key terms, persons, places, and concepts.

Online 99 Cents

http://www.epub14.51.19910.17.cram101.com/

Use www.Cram101.com for all your study needs

including Cram101's online interactive problem solving labs in chemistry, statistics, mathematics, and more.

CHAPTER OUTLINE: KEY TERMS, PEOPLE, PLACES, CONCEPTS
Chapter 18
Chemical reactions 2: reaction mechanisms driving the chemistry of life

- Transition state
- Carbocation
- Leaving group
- Nucleophile
- Nucleophilic substitution
- Racemic mixture
- Substitution reaction
- DNA-binding protein
- Electrophilic substitution
- Hydrolysis
- Avogadro constant
- Alkylation
- Carcinogen
- Gene expression
- Mutation
- DNA profiling
- Addition reaction
- Bioconjugation
- Electronegativity

Chapter 18. Chemical reactions 2: reaction mechanisms driving the chemistry of life

	Cyanohydrin
	Lewis structure
	Hydrogen cyanide
	Linamarin
	Amino acid
	Condensation
	Condensation reaction
	Dehydration reaction
	Polymerization
	Nucleic acid
	Peptide bond
	Phosphoric acid
	Salicylic acid
	Glucose 6-phosphate
	Hexokinase
	Phosphodiester bond
	Respiration
	Fructose 6-phosphate
	Glycolysis

Chapter 18. Chemical reactions 2: reaction mechanisms driving the chemistry of life

	Metabolism
	Fructose 1,6-bisphosphate
	Intramolecular force
	Keto-enol tautomerism
	Glyceraldehyde 3-phosphate
	Lactate dehydrogenase
	Lactic acid
	Oxidative phosphorylation
	Substrate-level phosphorylation
	Enolase
	Pyruvate dehydrogenase
	Citric acid
	Citric acid cycle
	Coenzyme A
	ATP synthase
	Electron transport chain
	Mechanism of action
	Atomic number
	Diameter

CHAPTER HIGHLIGHTS: KEY TERMS, PEOPLE, PLACES, CONCEPTS
Chapter 18. Chemical reactions 2: reaction mechanisms driving the chemistry of life

Transition state	The transition state of a chemical reaction is a particular configuration along the reaction coordinate. It is defined as the state corresponding to the highest energy along this reaction coordinate. At this point, assuming a perfectly irreversible reaction, colliding reactant molecules will always go on to form products.
Carbocation	A carbocation is an ion with a positively-charged carbon atom. The charged carbon atom in a carbocation is a 'sextet', i.e. it has only six electrons in its outer valence shell instead of the eight valence electrons that ensures maximum stability (octet rule). Therefore carbocations are often reactive, seeking to fill the octet of valence electrons as well as regain a neutral charge.
Leaving group	In chemistry, a leaving group is a molecular fragment that departs with a pair of electrons in heterolytic bond cleavage. Leaving groups can be anions or neutral molecules. Common anionic leaving groups are halides such as Cl^-, Br^-, and I^-, and sulfonate esters, such as para-toluenesulfonate ('tosylate', TsO^-).
Nucleophile	A nucleophile is a species that donates an electron-pair to an electrophile to form a chemical bond in a reaction. All molecules or ions with a free pair of electrons or at least one pi bond can act as nucleophiles. Because nucleophiles donate electrons, they are by definition Lewis bases.
Nucleophilic substitution	In organic and inorganic chemistry, nucleophilic substitution is a fundamental class of reactions in which an electron nucleophile selectively bonds with or attacks the positive or partially positive charge of an atom or a group of atoms called the leaving group; the positive or partially positive atom is referred to as an electrophile. The most general form for the reaction may be given as Nuc: + R-LG → R-Nuc + LG: The electron pair (:) from the nucleophile (Nuc) attacks the substrate (R-LG) forming a new bond, while the leaving group (LG) departs with an electron pair. The principal product in this case is R-Nuc.

Chapter 18. Chemical reactions 2: reaction mechanisms driving the chemistry of life

Racemic mixture	In chemistry, a racemic mixture, is one that has equal amounts of left- and right-handed enantiomers of a chiral molecule. The first known racemic mixture was 'racemic acid', which Louis Pasteur found to be a mixture of the two enantiomeric isomers of tartaric acid. Nomenclature A racemic mixture is denoted by the prefix (±)- or dl- (for sugars the prefix DL- may be used), indicating an equal (1:1) mixture of dextro and levo isomers.
Substitution reaction	In a substitution reaction, a functional group in a particular chemical compound is replaced by another group. In organic chemistry, the electrophilic and nucleophilic substitution reactions are of prime importance. Organic substitution reactions are classified in several main organic reaction types depending on whether the reagent that brings about the substitution is considered an electrophile or a nucleophile, whether a reactive intermediate involved in the reaction is a carbocation, a carbanion or a free radical or whether the substrate is aliphatic or aromatic.
DNA-binding protein	DNA-binding proteins are proteins that are composed of DNA-binding domains and thus have a specific or general affinity for either single or double stranded DNA. Sequence-specific DNA-binding proteins generally interact with the major groove of B-DNA, because it exposes more functional groups that identify a base pair. However there are some known minor groove DNA-binding ligands such as Netropsin, Distamycin, Hoechst 33258, Pentamidine and others. Examples DNA-binding proteins include transcription factors which modulate the process of transcription, various polymerases, nucleases which cleave DNA molecules, and histones which are involved in chromosome packaging and transcription in the cell nucleus.
Electrophilic substitution	Electrophilic substitution reactions are chemical reactions in which an electrophile displaces a group in a compound, typically but not always hydrogen. Electrophilic aromatic substitution is characteristic of aromatic compounds and is an important way of introducing functional groups onto benzene rings. The other main reaction type is electrophilic aliphatic substitution.
Hydrolysis	Hydrolysis usually means the rupture of chemical bonds by the addition of water. Generally, hydrolysis is a step in the degradation of a substance. In terms of the word's derivation, hydrolysis comes from Greek roots hydro 'water' + lysis 'separation'.

Chapter 18. Chemical reactions 2: reaction mechanisms driving the chemistry of life

Avogadro constant	In chemistry and physics, the Avogadro constant is defined as the ratio of the number of constituent particles N (usually atoms or molecules) in a sample to the amount of substance n (unit mole) through the relationship $N_A = Nn$. Thus, it is the proportionality factor that relates the molar mass of an entity, i.e., the mass per amount of substance, to the mass of said entity. The Avogadro constant expresses the number of elementary entities per mole of substance and it has the value $6.022\ 141\ 29(27) \times 10^{23}\ mol^{-1}$.
Alkylation	Alkylation is the transfer of an alkyl group from one molecule to another. The alkyl group may be transferred as an alkyl carbocation, a free radical, a carbanion or a carbene. Alkylating agents are widely used in chemistry because the alkyl group is probably the most common group encountered in organic molecules.
Carcinogen	A carcinogen is any substance, radionuclide, or radiation that is an agent directly involved in causing cancer. This may be due to the ability to damage the genome or to the disruption of cellular metabolic processes. Several radioactive substances are considered carcinogens, but their carcinogenic activity is attributed to the radiation, for example gamma rays and alpha particles, which they emit.
Gene expression	Gene expression is the process by which information from a gene is used in the synthesis of a functional gene product. These products are often proteins, but in non-protein coding genes such as ribosomal RNA (rRNA), transfer RNA (tRNA) or small nuclear RNA (snRNA) genes, the product is a functional RNA. The process of gene expression is used by all known life - eukaryotes (including multicellular organisms), prokaryotes (bacteria and archaea), possibly induced by viruses - to generate the macromolecular machinery for life. Several steps in the gene expression process may be modulated, including the transcription, RNA splicing, translation, and post-translational modification of a protein.
Mutation	In molecular biology and genetics, mutations are changes in a genomic sequence: the DNA sequence of a cell's genome or the DNA or RNA sequence of a virus. These random sequences can be defined as sudden and spontaneous changes in the cell. Mutations are caused by radiation, viruses, transposons and mutagenic chemicals, as well as errors that occur during meiosis or DNA replication.
DNA profiling	DNA profiling is a technique employed by forensic scientists to assist in the identification of individuals by their respective DNA profiles. DNA profiles are encrypted sets of numbers that reflect a person's DNA makeup, which can also be used as the person's identifier. DNA profiling should not be confused with full genome sequencing.

Chapter 18. Chemical reactions 2: reaction mechanisms driving the chemistry of life

Addition reaction	An addition reaction, in organic chemistry, is in its simplest terms an organic reaction where two or more molecules combine to form a larger one.
	Addition reactions are limited to chemical compounds that have multiple bonds, such as molecules with carbon-carbon double bonds (alkenes), or with triple bonds (alkynes). Molecules containing carbon--hetero double bonds like carbonyl (C=O) groups, or imine (C=N) groups, can undergo addition as they too have double bond character.
Bioconjugation	Bioconjugation is the process of coupling two biomolecules together in a covalent linkage. Common types of bioconjugation chemistry are amine coupling of lysine amino acid residues (typically through amine-reactive succinimidyl esters), sulfhydryl coupling of cysteine residues (via a sulfhydryl-reactive maleimide), and photochemically initiated free radical reactions, which have broader reactivity. The product of a bioconjugation reaction is a bioconjugate.
Electronegativity	Electronegativity, symbol χ, is a chemical property that describes the tendency of an atom or a functional group to attract electrons towards itself. An atom's electronegativity is affected by both its atomic number and the distance that its valence electrons reside from the charged nucleus. The higher the associated electronegativity number, the more an element or compound attracts electrons towards it.
Cyanohydrin	A cyanohydrin is a functional group found in organic compounds. Cyanohydrins have the formula $R_2C(OH)CN$, where R is H, alkyl, or aryl. Cyanohydrins are industrially important precursors to carboxylic acids and some amino acids.
Lewis structure	Lewis structures (also known as Lewis dot diagrams, electron dot diagrams, and electron dot structures) are diagrams that show the bonding between atoms of a molecule and the lone pairs of electrons that may exist in the molecule.
Hydrogen cyanide	Hydrogen cyanide is a chemical compound with chemical formula HCN. It is a colorless, extremely poisonous liquid that boils slightly above room temperature at 26 °C (79 °F). Hydrogen cyanide is a linear molecule, with a triple bond between carbon and nitrogen. A minor tautomer of HCN is HNC, hydrogen isocyanide.

Chapter 18. Chemical reactions 2: reaction mechanisms driving the chemistry of life

Linamarin	Linamarin is a cyanogenic glucoside found in the leaves and roots of plants such as cassava, lima beans, and flax. It is a glucoside of acetone cyanohydrin. Upon exposure to enzymes and gut flora in the human intestine, linamarin and its methylated relative lotaustralin can decompose to the toxic chemical hydrogen cyanide; hence food uses of plants that contain significant quantities of linamarin require extensive preparation and detoxification.
Amino acid	Amino acids (?'mi?no?, ?'ma?o?, or 'æm?o?) are molecules containing an amine group, a carboxylic acid group, and a side-chain that is specific to each amino acid. The key elements of an amino acid are carbon, hydrogen, oxygen, and nitrogen. They are particularly important in biochemistry, where the term usually refers to alpha-amino acids.
Condensation	Condensation is the change of the physical state of matter from gaseous phase into liquid phase, and is the reverse of vaporization. When the transition happens from the gaseous phase into the solid phase directly, the change is called deposition. Condensation is initiated by the formation of atomic/molecular clusters of that species within its gaseous volume--like rain drop or snow-flake formation within clouds--or at the contact between such gaseous phase and a (solvent) liquid or solid surface.
Condensation reaction	A condensation reaction is a chemical reaction in which two molecules or moieties (functional groups) combine to form one single molecule, together with the loss of a small molecule. When this small molecule is water, it is known as a dehydration reaction; other possible small molecules lost are hydrogen chloride, methanol, or acetic acid. The word 'condensation' suggests a process in which two or more things are brought 'together' to form something 'dense', like in condensation from gaseous to liquid state of matter; this does not imply, however, that condensation reaction products have greater density than reactants.
Dehydration reaction	In chemistry and the biological sciences, a dehydration reaction is usually defined as a chemical reaction that involves the loss of water from the reacting molecule. Dehydration reactions are a subset of elimination reactions. Because the hydroxyl group (-OH) is a poor leaving group, having a Brønsted acid catalyst often helps by protonating the hydroxyl group to give the better leaving group, $-OH_2^+$.
Polymerization	In polymer chemistry, polymerization is a process of reacting monomer molecules together in a chemical reaction to form polymer chains or three-dimensional networks. There are many forms of polymerization and different systems exist to categorize them.

Chapter 18. Chemical reactions 2: reaction mechanisms driving the chemistry of life

	In chemical compounds, polymerization occurs via a variety of reaction mechanisms that vary in complexity due to functional groups present in reacting compounds and their inherent steric effects explained by VSEPR Theory.
Nucleic acid	Nucleic acids are biological molecules essential for known forms of life on this planet; they include DNA (deoxyribonucleic acid) and RNA (ribonucleic acid). Together with proteins, nucleic acids are the most important biological macromolecules; each is found in abundance in all living things, where they function in encoding, transmitting and expressing genetic information.
	Nucleic acids were discovered by Friedrich Miescher in 1869. Experimental studies of nucleic acids constitute a major part of modern biological and medical research, and form a foundation for genome and forensic science, as well as the biotechnology and pharmaceutical industries.
Peptide bond	A peptide bond is a covalent chemical bond formed between two molecules when the carboxyl group of one molecule reacts with the amino group of the other molecule, causing the release of a molecule of water (H_2O), hence the process is a dehydration synthesis reaction (also known as a condensation reaction), and usually occurs between amino acids. The resulting C(O)NH bond is called a peptide bond, and the resulting molecule is an amide. The four-atom functional group -C(=O)NH- is called a peptide link.
Phosphoric acid	Phosphoric acid, is a mineral (inorganic) acid having the chemical formula H_3PO_4. Orthophosphoric acid molecules can combine with themselves to form a variety of compounds which are also referred to as phosphoric acids, but in a more general way. The term phosphoric acid can also refer to a chemical or reagent consisting of phosphoric acids, usually orthophosphoric acid.
Salicylic acid	Salicylic acid is a monohydroxybenzoic acid, a type of phenolic acid and a beta hydroxy acid. This colorless crystalline organic acid is widely used in organic synthesis and functions as a plant hormone. It is derived from the metabolism of salicin.
Glucose 6-phosphate	Glucose 6-phosphate is glucose sugar phosphorylated on carbon 6. This compound is very common in cells as the vast majority of glucose entering a cell will become phosphorylated in this way.

Chapter 18. Chemical reactions 2: reaction mechanisms driving the chemistry of life

Because of its prominent position in cellular chemistry, glucose 6-phosphate has many possible fates within the cell. It lies at the start of two major metabolic pathways:

- Glycolysis
- Pentose phosphate pathway

In addition to these metabolic pathways, glucose 6-phosphate may also be converted to glycogen or starch for storage.

Hexokinase

A hexokinase is an enzyme that phosphorylates a six-carbon sugar, a hexose, to a hexose phosphate. In most tissues and organisms, glucose is the most important substrate of hexokinases, and glucose-6-phosphate the most important product.

Variation across species

Genes that encode hexokinase have been discovered in each domain of life, ranging from bacteria, yeast, and plants to humans and other vertebrates.

Phosphodiester bond

A phosphodiester bond is a group of strong covalent bonds between a phosphate group and two 5-carbon ring carbohydrates (pentoses) over two ester bonds. Phosphodiester bonds are central to all known life, as they make up the backbone of each helical strand of DNA. In DNA and RNA, the phosphodiester bond is the linkage between the 3' carbon atom of one sugar molecule and the 5' carbon atom of another; the sugar molecules being deoxyribose in DNA and ribose in RNA.

The phosphate groups in the phosphodiester bond are negatively-charged. Because the phosphate groups have a pK_a near 0, they are negatively-charged at pH 7. This repulsion forces the phosphates to take opposite sides of the DNA strands and is neutralized by proteins (histones), metal ions such as magnesium, and polyamines.

Chapter 18. Chemical reactions 2: reaction mechanisms driving the chemistry of life

Respiration	In physiology, respiration (often confused with breathing) is defined as the transport of oxygen from the outside air to the cells within tissues, and the transport of carbon dioxide in the opposite direction. This is in contrast to the biochemical definition of respiration, which refers to cellular respiration: the metabolic process by which an organism obtains energy by reacting oxygen with glucose to give water, carbon dioxide and ATP (energy). Although physiologic respiration is necessary to sustain cellular respiration and thus life in animals, the processes are distinct: cellular respiration takes place in individual cells of the organism, while physiologic respiration concerns the bulk flow and transport of metabolites between the organism and the external environment.
Fructose 6-phosphate	Fructose 6-phosphate is fructose sugar phosphorylated on carbon 6 (i.e., is a fructosephosphate). The β-D-form of this compound is very common in cells. The vast majority of glucose and fructose entering a cell will become converted to this at some point.
Glycolysis	Glycolysis is the metabolic pathway that converts glucose $C_6H_{12}O_6$, into pyruvate, CH_3COCOO^- + H^+. The free energy released in this process is used to form the high-energy compounds ATP (adenosine triphosphate), $FADH_2$ and NADH (reduced nicotinamide adenine dinucleotide). Glycolysis is a definite sequence of ten reactions involving ten intermediate compounds (one of the steps involves two intermediates).
Metabolism	Metabolism is the set of chemical reactions that happen in the cells of living organisms to sustain life. These processes allow organisms to grow and reproduce, maintain their structures, and respond to their environments. The word metabolism can also refer to all chemical reactions that occur in living organisms, including digestion and the transport of substances into and between different cells, in which case the set of reactions within the cells is called intermediary metabolism or intermediate metabolism.
Fructose 1,6-bisphosphate	Fructose 1,6-bisphosphate is fructose sugar phosphorylated on carbons 1 and 6 (i.e., is a fructosephosphate). The β-D-form of this compound is very common in cells. The vast majority of glucose and fructose entering a cell will become converted to fructose 1,6-biphosphate at some point.
Intramolecular force	An intramolecular force is any force that holds together the atoms making up a molecule or compound. They contain all types of chemical bond. They are stronger than intermolecular forces, which are present between atoms or molecules that are not actually bonded.

Chapter 18. Chemical reactions 2: reaction mechanisms driving the chemistry of life

Keto-enol tautomerism	In organic chemistry, keto-enol tautomerism refers to a chemical equilibrium between a keto form (a ketone or an aldehyde) and an enol (an alcohol). The enol and keto forms are said to be tautomers of each other. The interconversion of the two forms involves the movement of a proton and the shifting of bonding electrons; hence, the isomerism qualifies as tautomerism.
Glyceraldehyde 3-phosphate	Glyceraldehyde 3-phosphate, GADP, GAP, TP, GALP or PGAL, is a chemical compound that occurs as an intermediate in several central metabolic pathways of all organisms. It is a phosphate ester of the 3-carbon sugar glyceraldehyde and has chemical formula $C_3H_7O_6P$. The CAS number of glyceraldehyde 3-phosphate is 591-59-3 and that of D-glyceraldehyde 3-phosphate is 591-57-1. An intermediate in both glycolysis and gluconeogenesis Formation D-glyceraldehyde 3-phosphate is formed from the following three compounds in reversible reactions: • Fructose-1,6-bisphosphate (F1,6BP), catalyzed by aldolase. Compound C05378 at KEGG Pathway Database. Enzyme 4.1.2.13 at KEGG Pathway Database. Compound C00111 at KEGG Pathway Database. Compound C00118 at KEGG Pathway Database. The numbering of the carbon atoms indicates the fate of the carbons according to their position in fructose 6-phosphate.
Lactate dehydrogenase	D-lactate dehydrogenase, membrane binding Lactate dehydrogenase is an enzyme (EC 1.1.1.27) present in a wide variety of organisms, including plants and animals. Lactate dehydrogenases exist in four distinct enzyme classes. Two of them are cytochrome c-dependent enzymes, each acting on either D-lactate (EC 1.1.2.4) or L-lactate (EC 1.1.2.3).

Chapter 18. Chemical reactions 2: reaction mechanisms driving the chemistry of life

Lactic acid	Lactic acid, is a chemical compound that plays a role in various biochemical processes and was first isolated in 1780 by the Swedish chemist Carl Wilhelm Scheele. Lactic acid is a carboxylic acid with the chemical formula $C_3H_6O_3$. It has a hydroxyl group adjacent to the carboxyl group, making it an alpha hydroxy acid (AHA).
Oxidative phosphorylation	Oxidative phosphorylation is a metabolic pathway that uses energy released by the oxidation of nutrients to produce adenosine triphosphate (ATP). Although the many forms of life on earth use a range of different nutrients, almost all aerobic organisms carry out oxidative phosphorylation to produce ATP, the molecule that supplies energy to metabolism. This pathway is probably so pervasive because it is a highly efficient way of releasing energy, compared to alternative fermentation processes such as anaerobic glycolysis.
Substrate-level phosphorylation	Substrate-level phosphorylation is a type of metabolism that results in the formation and creation of adenosine triphosphate (ATP) or guanosine triphosphate (GTP) by the direct transfer and donation of a phosphoryl (PO_3) group to adenosine diphosphate (ADP) or guanosine diphosphate (GDP) from a phosphorylated reactive intermediate. Note that the phosphate group does not have to directly come from the substrate. By convention, the phosphoryl group that is transferred is referred to as a phosphate group.
Enolase	Enolase is a metalloenzyme responsible for the catalysis of the conversion of 2-phosphoglycerate (2-PG) to phosphoenolpyruvate (PEP), the ninth and penultimate step of glycolysis. Enolase can also catalyze the reverse reaction, depending on environmental concentrations of substrates. The optimum pH for this enzyme is 6.5. Enolase is present in all tissues and organisms capable of glycolysis or fermentation. The enzyme was discovered by Lohmann and Meyerhof in 1934, and has since been isolated from a variety of sources including human muscle and erythrocytes.
Pyruvate dehydrogenase	Pyruvate dehydrogenase is the first component enzyme of pyruvate dehydrogenase complex (PDC). The pyruvate dehydrogenase complex contributes to transforming pyruvate into acetyl-CoA by a process called pyruvate decarboxylation. Acetyl-CoA may then be used in the citric acid cycle to carry out cellular respiration, so pyruvate dehydrogenase contributes to linking the glycolysis metabolic pathway to the citric acid cycle and releasing energy via NADH. EC 1.2.4.1.
Citric acid	Citric acid is a weak organic acid. It is a natural preservative/conservative and is also used to add an acidic, or sour, taste to foods and soft drinks. In biochemistry, the conjugate base of citric acid, citrate, is important as an intermediate in the citric acid cycle, and therefore occurs in the metabolism of virtually all living things.

Chapter 18. Chemical reactions 2: reaction mechanisms driving the chemistry of life

Citric acid cycle	The citric acid cycle -- also known as the tricarboxylic acid cycle (TCA cycle), the Krebs cycle, or the Szent-Györgyi-Krebs cycle -- is a series of chemical reactions used by all aerobic organisms to generate energy through the oxidization of acetate derived from carbohydrates, fats and proteins into carbon dioxide and water. In addition, the cycle provides precursors for the biosynthesis of compounds including certain amino acids as well as the reducing agent NADH that is used in numerous biochemical reactions. Its central importance to many biochemical pathways suggests that it was one of the earliest established components of cellular metabolism and may have originated abiogenically.
Coenzyme A	Coenzyme A is a coenzyme, notable for its role in the synthesis and oxidation of fatty acids, and the oxidation of pyruvate in the citric acid cycle. All sequenced genomes encode enzymes that use coenzyme A as a substrate, and around 4% of cellular enzymes use it as a substrate. It is adapted from cysteamine, pantothenate, and adenosine triphosphate.
ATP synthase	ATP synthase is an important enzyme that provides energy for the cell to use through the synthesis of adenosine triphosphate (ATP). ATP is the most commonly used 'energy currency' of cells from most organisms. It is formed from adenosine diphosphate (ADP) and inorganic phosphate (P_i), and needs energy.
Electron transport chain	An electron transport chain couples electron transfer between an electron donor (such as NADH) and an electron acceptor (such as O_2) with the transfer of H^+ ions (protons) across a membrane. The resulting electrochemical proton gradient is used to generate chemical energy in the form of adenosine triphosphate (ATP). Electron transport chains are the cellular mechanisms used for extracting energy from sunlight in photosynthesis and also from redox reactions, such as the oxidation of sugars (respiration).
Mechanism of action	In pharmacology, the term mechanism of action refers to the specific biochemical interaction through which a drug substance produces its pharmacological effect. A mechanism of action usually includes mention of the specific molecular targets to which the drug binds, such as an enzyme or receptor. For example, the mechanism of action of aspirin involves irreversible inhibition of the enzyme cyclooxygenase, therefore suppressing the production of prostaglandins and thromboxanes, thereby reducing pain and inflammation.

Chapter 18. Chemical reactions 2: reaction mechanisms driving the chemistry of life

Atomic number	In chemistry and physics, the atomic number is the number of protons found in the nucleus of an atom and therefore identical to the charge number of the nucleus. It is conventionally represented by the symbol Z. The atomic number uniquely identifies a chemical element. In an atom of neutral charge, the atomic number is also equal to the number of electrons.
Diameter	In geometry, a diameter of a circle is any straight line segment that passes through the center of the circle and whose endpoints are on the circle. The diameters are the longest chords of the circle. The word 'diameter' derives from Greek δι?μετρος (diametros), 'diagonal of a circle', from δια- (dia-), 'across, through' + μ?τρον (metron), 'a measure').

PRACTICE QUIZ
Chapter 18. Chemical reactions 2: reaction mechanisms driving the chemistry of life

1. In a _____, a functional group in a particular chemical compound is replaced by another group. In organic chemistry, the electrophilic and nucleophilic _____s are of prime importance. Organic _____s are classified in several main organic reaction types depending on whether the reagent that brings about the substitution is considered an electrophile or a nucleophile, whether a reactive intermediate involved in the reaction is a carbocation, a carbanion or a free radical or whether the substrate is aliphatic or aromatic.

    a. Substrate
    b. Telomerization
    c. Thermal decomposition
    d. Substitution reaction

2. In polymer chemistry, _____ is a process of reacting monomer molecules together in a chemical reaction to form polymer chains or three-dimensional networks. There are many forms of _____ and different systems exist to categorize them.

    In chemical compounds, _____ occurs via a variety of reaction mechanisms that vary in complexity due to functional groups present in reacting compounds and their inherent steric effects explained by VSEPR Theory.

    a. Bulk polymerization
    b. Cationic polymerization
    c. Polymerization
    d. Living anionic polymerization

3. D-_____, membrane binding _____ is an enzyme (EC 1.1.1.27) present in a wide variety of organisms, including plants and animals.

    _____s exist in four distinct enzyme classes. Two of them are cytochrome c-dependent enzymes, each acting on either D-lactate (EC 1.1.2.4) or L-lactate (EC 1.1.2.3).

    a. Lactate dehydrogenase
    b. Medullary thyroid cancer
    c. Neprilysin
    d. NMP22

---

4. _____s are proteins that are composed of DNA-binding domains and thus have a specific or general affinity for either single or double stranded DNA. Sequence-specific _____s generally interact with the major groove of B-DNA, because it exposes more functional groups that identify a base pair. However there are some known minor groove DNA-binding ligands such as Netropsin, Distamycin, Hoechst 33258, Pentamidine and others.

    Examples
    _____s include transcription factors which modulate the process of transcription, various polymerases, nucleases which cleave DNA molecules, and histones which are involved in chromosome packaging and transcription in the cell nucleus.

    a. DNA-binding protein
    b. Fluorescence recovery after photobleaching
    c. FlyFactorSurvey
    d. GHK flux equation

5. In organic and inorganic chemistry, _____ is a fundamental class of reactions in which an electron nucleophile selectively bonds with or attacks the positive or partially positive charge of an atom or a group of atoms called the leaving group; the positive or partially positive atom is referred to as an electrophile.

    The most general form for the reaction may be given as

    Nuc: + R-LG → R-Nuc + LG:

    The electron pair (:) from the nucleophile (Nuc) attacks the substrate (R-LG) forming a new bond, while the leaving group (LG) departs with an electron pair. The principal product in this case is R-Nuc.

    a. Partial oxidation
    b. Photodegradation
    c. Nucleophilic substitution
    d. Photoinduced electron transfer

ANSWER KEY
Chapter 18. Chemical reactions 2: reaction mechanisms driving the chemistry of life

1. d
2. c
3. a
4. a
5. c

You can take the complete Chapter Practice Test

for Chapter 18. Chemical reactions 2: reaction mechanisms driving the chemistry of life
on all key terms, persons, places, and concepts.

Online 99 Cents

http://www.epub14.51.19910.18.cram101.com/

Use www.Cram101.com for all your study needs

including Cram101's online interactive problem solving labs in chemistry, statistics, mathematics, and more.

Other Cram101 e-Books and Tests

Want More?
Cram101.com...

Cram101.com provides the outlines and highlights of your textbooks, just like this e-StudyGuide, but also gives you the PRACTICE TESTS, and other exclusive study tools for all of your textbooks.

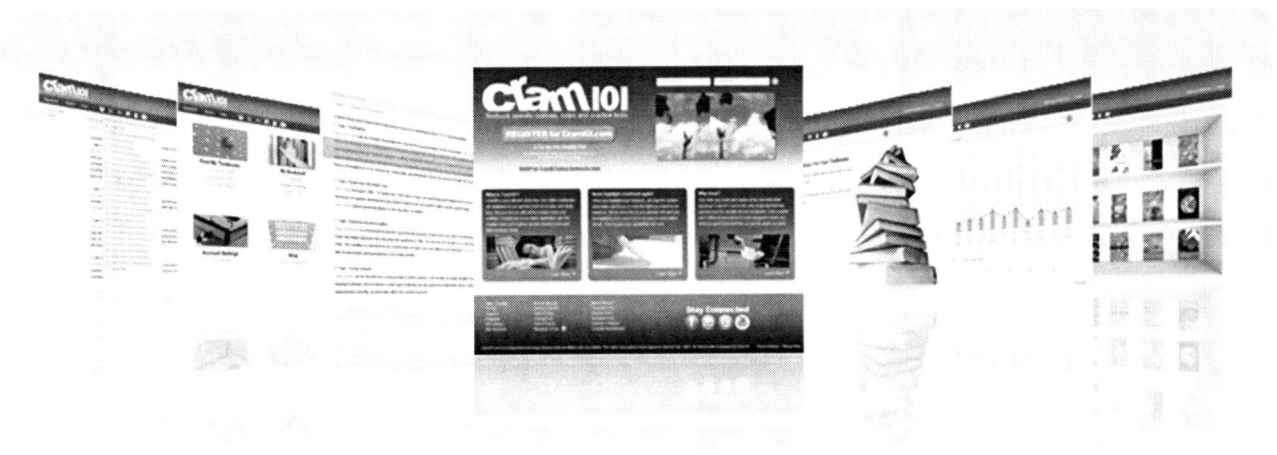

Learn More. *Just click*
http://www.cram101.com/

Other Cram101 e-Books and Tests

Want More?
Cram101.com...

Cram101.com provides the outlines and highlights of your textbooks, just like this e-StudyGuide, but also gives you the PRACTICE TESTS, and other exclusive study tools for all of your textbooks.

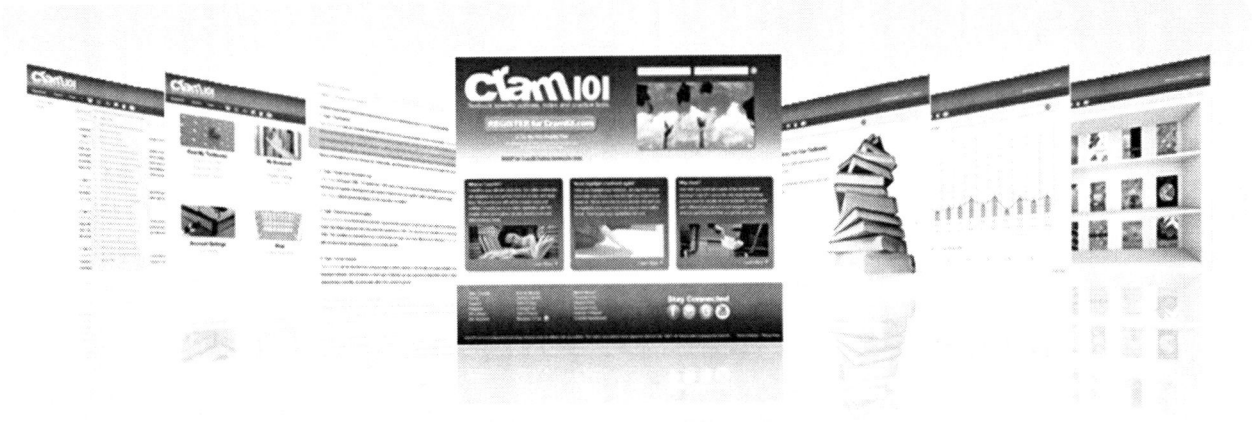

Learn More. *Just click*
http://www.cram101.com/

Other Cram101 e-Books and Tests